Finance and Innovation

D W Budworth
Associate Fellow,
CENTRIM,
University of Brighton

INTERNATIONAL THOMSON BUSINESS PRESS
I ⓣ P An International Thomson Publishing Company

London • Bonn • Boston • Johannesburg • Madrid • Melbourne • Mexico City • New York • Paris
Singapore • Tokyo • Toronto • Albany, NY • Belmont, CA • Cincinnati, OH • Detroit, MI

Finance and Innovation

Copyright © 1996 D. W. Budworth

First published 1996 by International Thomson Business Press

 I(T)P   A division of International Thomson Publishing Inc.
The ITP logo is a trademark under licence

British Library Cataloguing-in-Publication Data
A catalogue record for this book is available from the British
Library

First edition 1996

Typeset by WestKey Limited, Falmouth, Cornwall
Printed in the UK by T.J. Press (Padstow) Ltd., Padstow,
Cornwall

ISBN 0-415-13258-4

International Thomson Business Press
Berkshire House
168–173 High Holborn
London WC1V 7AA
UK

International Thomson Business Press
20 Park Plaza
14th Floor
Boston MA 02116
USA

http://www.thomson.com/itbp.html

In affectionate memory of

Duncan Davies,

pioneer of technological economics,

intellectual iconoclast and stimulator,

in gratitude for his tireless contributions

to many a common cause

Contents

Figures

Tables

Preface

Innovation is at last beginning to attract the attention which its importance deserves, but it is still not regarded as the integral and vital part of industrial and commercial operation which it undoubtedly is. This lack of integration is particularly marked in relation to finance.

When the subject of finance and innovation is considered, it is most usually in relation to raising finance for a new technology-based venture. While important, this is in my view only a small part of the general problem, and in any case one in which most of the difficulties are common to any new venture, whether particularly innovative or not.

Although, over the past twelve years or so, I have attempted from time to time to develop an approach which incorporates innovation as an integral part of the financial operation of a company, the idea of writing a book on the broader subject of finance and innovation did not occur to me until John Bessant invited me to consider the possibility. In view of my varied background and established interests and the lack of any other works on the subject, the challenge was too great to refuse.

The focus of the book is on the firm and its strategic management, not on the management of the innovation process itself. Issues such as life cycles, experience curves, and the broader topic of project selection and management in which they fall are well served by the existing literature, such as Twiss (1992), and are mentioned only in passing. The final chapter deals with some aspects of macroeconomics and national policy as they affect the firm, but again makes no attempt to give a full account of a topic well covered by the references given in it.

My aim has been to explain the principles, and the main themes on which I have attempted to string their presentation are the importance of cash and the model of a company or business as a series of innovations. This model provides a framework for considering a company as an integrated, dynamic, living system, relying for its survival principally on its intellectual assets.

If the book appears to concentrate too much on finance at the expense of innovation, that is because I believe that finance provides the framework within which innovation must take place, and that accounting provides the basic language in which dialogue can be carried out. Accountants do not merely count the beans: they define what counts as a bean. Further, the extent of dissatisfaction in the accounting world and the scope for constructive input to the debates which are currently taking place within it are not appreciated outside the narrow circle of professionals and a few others who have taken an interest in the issues. The apparent precision of accountancy tends to obscure – especially to those trained in the exact sciences – its comparative lack of accuracy, and I hope that any misconceptions on this point will not survive a reading of this book.

I have focused largely on the UK, with some reference to the US, especially where the thinking in that country is ahead of ours and its influence strong. The situations in the UK and the US are similar, particularly in comparison with those prevailing in other countries such as Germany and Japan, but the UK is even more dependent than the US on institutional finance. Some reference to political and historical factors has been made where these are needed to explain developments.

The worlds of finance and innovation in the UK too often regard each other with mutual incomprehension, sometimes bordering on hostility or contempt. The need to develop means of communication between two very different ways of looking at the industrial world has been underlined by several recent publications, echoing earlier policy pronouncements going back for many years. I hope that this book will make some contribution to that end.

London
August 1995

Acknowledgements

Many of those with whom I have come into contact in the course of a somewhat varied career have influenced the thinking which has gone into this book, and it would be impossible to list them all. My views on the critical importance of company size in innovation were influenced in the 1970s by Sir Austin Bide of Glaxo and Sir Denning Pearson of Rolls-Royce; Dr F E Jones of Mullard introduced me to the virtues of the value added concept at around the same time, while a decade later Geoff Smith inspired its incorporation into financial modelling, a field in which Dr Gordon Fryers of Reckitt and Colman was also influential. The statistical studies of Mrs Joan Cox provided material for models and support for some of their strategic implications. Henry Gold and Allan Cook, in their time at Shell, were most helpful in explaining the intricacies of accounting standards, and more recently I have had valuable insights from Professor John Grinyer of the University of Dundee and Michael Mumford of the University of Lancaster.

This book would not have been written if Professor John Bessant of the University of Brighton had not suggested it. Professor Michael Gibbons of the University of Sussex, Hugh J Osburn of American Appraisal (UK) Ltd, Dr Clive Phillips of Kobe Steel, and Dr William Nixon of the University of Dundee have all been most helpful in drawing my attention to information of which I was unaware, while Dr Brian Newbould formerly of ICI, Dr Ted Ellis of Pilkington, and Dr Richard Reeves of Cranfield University have been generous with advice and information. The resources of the British Library and of the City Business Library have been invaluable to one lacking an institutional base.

Finally, my thanks must go to my long-suffering wife, Kathleen who, despite knowing from previous experience what writing a book entails, nevertheless encouraged me to undertake another and patiently bore the consequences.

I am grateful to Airbus Industrie for supplying the data on which Figure 6.6 is based and for permission to use them; to the Policy Studies Institute for permission to reproduce Table 7.1; and to OECD for permission to reproduce Table 1.1. Extracts from 'The Corporate Report', SSAP 13, the ASB Statement of Aims, the ASB Statement of Principles, and FRS 5 are reproduced with permission from the Accounting Standards Board.

Chapter 1

Innovation and the company context

INNOVATION

Innovation: difficult but essential

The essence of innovation is novelty. Innovation is both the process of introducing something new and the new thing itself. It is thus a concept of very general application. Indeed, one of the best-known views of innovation is that of Machiavelli, the early sixteenth century writer on politics, whose view of constitutional innovation was that it was difficult, dangerous, and doubtful of success.

Machiavelli pointed out that the innovator makes enemies of those who prospered under the old order, but receives only lukewarm support from those who stand to benefit from the new. He attributed this half-hearted support to two causes – fear and incredulity, and further pointed out that power was necessary to bring about innovation: those who have to rely on persuasion seldom succeed, partly because while it is easy to persuade people of the desirability of change, it is difficult to generate enough commitment on their behalf to bring about that change in the face of inevitable difficulties.

This book is concerned with innovation in the economic context, where it refers to the introduction of new offerings to the market, usually by companies. The focus will be on innovation in the company context, rather than on its significance in the economy as a whole. In the company context, innovation may well be as difficult, uncertain, and dangerous as Machiavelli suggests, but the absence of innovation is in the long run certainly much more dangerous, albeit possibly less difficult in the short run: a company which fails to innovate will eventually find that its performance declines relative to its competitors. As a result, the company may well fail altogether. Failure to innovate in response to new circumstances is the main reason why the average life of a company tends to be rather less than that of the average human being.

A company which wants to survive must innovate: innovation is therefore a literally vital – essential to life – function within a company. To bring innovation about, contributions from many parts of a company's operations must be brought together: innovation is therefore an integral part of the company, not the sole responsibility of the research and development (R and D) department, the marketing department, nor of any other functional unit.

Neither of these attributes of innovation is as fully and universally recognised as it should be. Innovation is still often implicitly assumed to be an occasional, special, and even an optional, aspect of company life, rather than part of its normal activities. This assumption has particularly harmful effects when the financial implications of innovation are considered, since those responsible for financial matters tend not to regard innovation as vital, while those primarily engaged in innovative activity tend not to regard it as integral. It is the aim of this book to contribute to bridging this gap.

The innovative performance of a company is affected by the conditions of the economy in which it operates. These in turn are largely affected by the actions of the relevant government. Much economic study of innovation has therefore been undertaken with a view to affecting public policy, and the topic is well documented. For the most part, this book will take the economic and other boundary conditions under which a company operates as fixed, and will concentrate on the financial aspects of innovative performance which the company itself can affect.

Definitions of innovation

There are many definitions of innovation. The Economic and Social Research Council (ESRC) of the UK, which has supported a substantial amount of academic research on the economic and social aspects of innovation over the past two decades, selected innovation as its first priority theme under the new terms of reference which it was given in 1993. These terms of reference make explicit reference to meeting the needs of users of research, and contributing to the economic competitiveness of the UK, the effectiveness of public policy, and the quality of life. Under the clear influence of these terms of reference, the ESRC's definition of innovation is:

> The successful creation, development and application of new techniques or ways of working that improve the effectiveness and efficiency of individuals and organisations. It is a major driving force of wealth creation in a competitive market economy, and contributes thereby to improved quality of life.

This definition reflects that the role of innovation in economic growth has been a major subject of study in economics for some time. The conclusions and approaches of economists have influenced public policy, whereas the work of those sociologists who have studied the human aspects of innovation have probably had more influence on the theory and practice of management.

There is still, however, something of a gap between these academic studies and the day-to-day practice – or absence of practice – of innovation in companies. Stimulating innovation in industry is regarded by the UK government's Department of Trade and Industry (DTI) as its main role in the science and technology area, and since 1991 it has had an Innovation Unit, which defines innovation in this context very simply as:

The successful exploitation of new ideas.

which is the definition which underlies the approach of this book.

Innovation in the business context

Although the essence of innovation is new ideas, the ideas themselves do not need to be novel in an absolute sense, as long as they are new in the context in which they are exploited. A new application of an old idea may be commercially very successful and economically important, and will give rise to many of the same problems and issues as does the exploitation of a genuinely new idea.

The management writer Peter Drucker distinguishes seven sources of innovative opportunity (Drucker 1985): the unexpected, such as a change in customer taste becoming evident in the market; incongruities, such as between the assumption that the key to freight shipping costs was making ships more efficient when in motion, whereas the real key was reducing their costs in port; process need; industry and market structures, such as the globalisation of the automobile industry; demographics, which change the age structure of the population and hence the demand for such things as disposable nappies, higher education, or sheltered retirement homes; changes in perception, especially about how individuals see themselves and their role as consumers; and new knowledge.

The key entrepreneurial contribution is to recognise these opportunities when they arise: except in the last case, following them up may be relatively straightforward, involving no more risks than with any new business. Such risks are not negligible, but they are well understood and can be readily assessed by any experienced business manager, bank manager, or investment institution.

Drucker points out that innovation based on new knowledge, which need not necessarily be scientific or technical, is the form of innovation which attracts most attention, gets most money, and ranks high in historical importance. It differs from the other forms of innovation in its timescale, failure rate, predictability, and difficulty. Drucker characterises new knowledge-based innovation as 'temperamental, capricious, and hard to manage', echoing the experience of Machiavelli once again. It is this type of innovation which causes most problems, not least financial problems, and which will therefore feature most strongly in this book.

Just as it is possible for innovation to take place without there being any contribution from new knowledge, particularly new technical knowledge, it is more than possible, and indeed quite frequent, for new technical ideas or capabilities to emerge without their giving rise to innovation. Such developments are either discoveries about how the world works or inventions of devices to produce some practical effect. To turn them into innovations there has to be a market opportunity as well as a technical possibility, and the road from the possibility to its successful exploitation is often long and hard.

Stages of innovation

Once economists had become convinced, around forty or fifty years ago, that innovation and technical change were responsible for the bulk of economic growth, they began to study the process in some detail, and they became particularly concerned to devise ways by which to measure the economic resources devoted to it. Perhaps unduly influenced by the experience of the Second World War, in which science was exploited for military innovations – most spectacularly the nuclear fission bomb – they envisaged a sequence of stages as follows:

1 *Basic research*, carried out, typically in universities, in order to find out how the physical world works, with no thought to, or apparent possibility of, practical application. At this stage the value of the results is completely unpredictable, but the costs are modest.

Pre-war nuclear physics was a good example of such esoteric studies, and the 1933 view of Ernest Rutherford (the creator of the field and its foremost experimentalist until his death in 1937), that the concept of obtaining power from nuclear transmutation was 'the merest moonshine' is often quoted. Shortly afterwards, however, the physicist Leo Szilard discerned the prospect of doing so through a self-sustaining chain reaction from which energy could be extracted. Although neither he nor anyone else had discovered which materials, if any, had the necessary properties, he made a patent application for the idea in 1934. Recognising its military potential, he assigned the patent to the British Admiralty in order to keep it secret. It was not until early 1939 that uranium was found to have the right characteristics to fulfil the conditions envisaged by Szilard's patent.

2 *Applied research*, carried out, usually by industrial or government laboratories, with a view to obtaining new knowledge on which to base the design of some means of exploiting the fundamental knowledge arising from basic research. At this stage the costs are usually higher because there is a good deal of ground to explore, but there is some possibility of a return.

In the nuclear bomb case, the main issue to explore was how large a mass of material would be needed to sustain the necessary chain reaction. Szilard's patent covered this concept of a critical mass.

3 *Development* or 'experimental development', in which the knowledge from applied research is used to design and test prototype products, processes, or services. The costs may be quite large at this stage because real artefacts are

being made, but the chances of a return are judged as being reasonably good.

In the nuclear bomb case, this stage was severely restricted by shortage of fissile material.

4 *Production and marketing,* often involving considerable expenditure on manufacturing facilities, training of the production workers, building up stocks, publicity, and so on. At this stage, the assessment has been made that the project will be profitable.

In the nuclear fission bomb case, this stage came after the end of the war. The marketing was straightforward, as the customer, the government, was also the producer.

An often-quoted rule of thumb is that for every 1 unit of expenditure on research, 10 units need to be spent on development, and 100 on production and marketing. The figures in this rule should not be taken too literally, but the basic point is valid – costs go up significantly at every stage.

There is very little reliable information about the distribution of expenditure between the various stages of innovation. Such evidence as there is comes from occasional surveys. The intergovernmental organisation OECD has collected some figures which are shown in Table 1.1.

Table 1.1 Expenditure at different stages of innovation (%)

Type of Expenditure	Germany	Italy	Finland
R and D	26.0	17.9	39.0
Design and engineering	22.0	25.2	6.3
Patents and licences	2.0	—	4.0
Start-up investment	46.0	51.5	46.2
Marketing	4.0	5.4	4.5

Source: © OECD, (1995), *National Systems for Financing Innovation.* Reproduced by permission of the OECD.

It is clear from the figures in Table 1.1 that the start-up investment does indeed contribute the major part of the expenditure on innovation, but the difficulty of distinguishing between the various stages, the complexity of real innovation, and the variation in its financial structure from one industry to another mean that the figures cannot be regarded as other than general guides.

Statistics of innovation

In order to understand the economic significance of the various processes, data on expenditure were required. The first comprehensive figures for the UK were those gathered by the FBI for 1945, of which further details are given in Chapter 7. Government statisticians took an interest in the subject after the war, as did those of other countries. In 1963, the leading industrial countries standardised the definitions for the collection of data through OECD. The OECD fully recognised that the innovation process consisted of many stages, but was able to reach agreement on the definitions only of basic research, applied research, and experimental development, essentially on the lines given above.

These definitions, and the limitation of the available statistics to those on research and development (R and D), have had a very profound influence in the years since 1963. Much economic and political discussion and analysis of innovation is, for lack of any other data, based on R and D figures which are themselves collected on the OECD definitions, known as the 'Frascati' definitions after the resort at which the meeting to agree them was held.

The non-R and D fields of innovative activity listed by OECD (1994) are:

1 *Tooling-up and industrial engineering*, covering costs to set up manufacture.
2 *Manufacturing start-up and pre-production development*, which may include product or process modification, retraining, and trial production.
3 *Marketing for new products*, covering market research, test marketing, adaptation of products for specific markets, and launch advertising. OECD (1994) excludes non-recurring costs of establishing distribution, maintenance, and sales channels, which were included in the previous version of its guidelines.
4 *Acquisition of disembodied technology*, such as patents and know-how, from external as well as internal sources.
5 *Acquisition of embodied technology*, such as machinery and equipment.
6 *Design*, including organisational as well as technological aspects.

Although expressed largely in the language appropriate to manufacturing, the concepts also apply to service innovations. The activities need not be sequential, being brought together by the techniques of 'simultaneous engineering', but the list nonetheless gives some indication of the complexity of the total innovation process and the range of expertise which has to be brought to bear on it.

Both the OECD and the European Commission (EC) have recently made attempts to measure some of these other aspects of innovation. As well as more comprehensive coverage of costs, these attempts include other inputs to innovation, such as the supply of trained people, sources of information, and technology transfer; and the outcomes of innovation, such as percentage of sales from new products, patents, technological balance of payments, and assessments of the impact of scientific publications. The issue of measurement of innovation outcomes is at present in an embryonic stage, and discussion of it in this book is deferred until Chapter 7.

Limitations of the linear model of innovation

The so-called 'linear model' of innovation, with its orderly progression from basic research to production and marketing and, by implication, to economic success, is a gross oversimplification of reality. Nevertheless, the separate stages can be distinguished in practice, particularly in hindsight, and especially in new industries based on new scientific discoveries, such as biotechnological exploitations of the discovery of DNA and how to manipulate it, and the nuclear industry already used as an example. Even in these cases, however, as soon as the exploition really gets under way, the relationship between the stages tends to become complex and iterative with, for example, new research needs being thrown up by development and exploitation, and applied research being steered by targets for exploitation.

The over-simplicity of the model is nonetheless dangerous because of its implicit policy recommendation that, if basic research is properly looked after, everything else, including economic success, will follow. This belief is dangerously mistaken. The policy was, however, followed for many years both by industry and government, with industry discovering its limitations first. Having failed to make a success of 'country house' laboratories, founded in the belief that employing scientists and leaving them free to work on whatever took their fancy would lay the foundations of future prosperity, industry's representatives became early critics of the linear model as a foundation for public policy (Budworth 1973). More recently, academic students of innovation have recognised the inadequacies of the linear model, although they have, so far, failed to come up with any universally-accepted framework for their studies and for the collection of data.

Types of innovation

Largely due to the work of academic students of innovation, innovation has been subdivided into categories by criteria of type and degree. These subdivisions can be useful for analysis but, like those of the stages of innovation, should be regarded as no more than aids to understanding, and not as representing fundamental distinctions.

The first division is into product and process innovations, which are fairly self-explanatory terms. The significance is that product innovations have economic effects on sales and output, whereas process innovations have effects on efficiency. In practice, the two are closely tied together. A large part of the Japanese success in the motor bicycle industry, for example, was due to process development (Boston Consulting Group 1975). A new product may require a new process before it can be made; while a new process will often open up opportunities for making products which were not economically feasible before.

A good example of the latter case is the development by the UK glassmaker Pilkington of the float glass process, discussed in more detail in Chapter 6. This process forms a ribbon of flat glass by pouring molten glass on to a bath of liquid tin where it cools with perfectly flat, parallel, and unblemished faces. The process was originally envisaged as replacing the plate glass process in which a rolled ribbon of glass was ground and then polished on both sides, but it proved to be cheap enough also to replace the sheet glass process in which a ribbon of glass was drawn vertically from the melt. It was also found to lend itself to modifying the glass surfaces by a variety of chemical and physical means, with the result that it could be exploited to produce a wide range of products with properties tailored to different requirements for optical and thermal transmission.

Although the development and introduction of the float glass process was difficult and expensive, one thought which sustained the company throughout it was that if it was successful the rewards would be great. The market for flat glass was large, established, and unlikely to disappear, so that all effort could be concentrated on making the process work. A product innovation, on the other hand, always has an element, sometimes large, of market uncertainty.

Degrees of innovation

Academic students of innovation distinguish between incremental, radical, and fundamental innovation. Again, these are somewhat arbitrary distinctions, although there is no doubt that innovations do differ in the amount of change which they bring about. These differences in turn mean that the financial demands and the risk of the innovation vary from case to case.

Incremental innovation is, as its name suggests, the type which makes only small changes at any one time. The cumulative effect over time may, however, be large. The internal combustion engine in a modern automobile is recognisably the descendant of the original version of just over a hundred years ago, but its efficiency is considerably better, as are its longevity and reliability. A whole series of incremental innovations in mechanical design and materials, backed up by studies of the details of its working, have enabled it to hold its own against the attack of many competing designs.

Incremental innovations are made by existing manufacturers and, by definition, do not involve major investments or risks. Incremental innovation may be the only type of innovation which is possible in some fields, a fact which increases its importance. In the process field, wine making and cement manufacture are examples of continuous incremental innovation over a long period.

Radical innovations make a substantial change, but stop short of founding a new industry. Float glass is a good example. The risks and required investment in radical innovations are usually considerably greater than for incremental innovation, and they offer more of an opportunity for new producers to enter the market.

Fundamental innovations are those which depend on major new scientific knowledge, and open up completely new industries. Nuclear power has already been mentioned, as has the new biotechnology. Another example, possibly the most economically-significant one for the late twentieth and early twenty-first centuries, is the transistor, which depended on electron movements in a solid, as opposed to the electron movements in a vacuum which characterised the electronic valve (or tube, in the US), and which led to the integrated circuit and the cheap and reliable computer.

Fundamental innovations are usually introduced by new firms. It is often possible to start on a relatively small scale since the technology is immature and the markets not established. The biotechnology sector is a good example with many small firms being started in the past ten or fifteen years. Once the viability and promise of new products are established, however, it may be necessary to operate on a much larger scale, either by being taken over by a larger firm, or by working closely with one.

The economic view

The distinction between invention and innovation was introduced into economic theory by Schumpeter (1911) in the early part of this century, but it has still not been entirely absorbed into economic or political thinking. To be economically really significant, innovation has to be widely adopted through the process which Schumpeter christened diffusion. Economists are therefore concerned with all three stages, the links between them, and the factors affecting them, not least of which is the supply of finance.

Invention, or discovery, can be the work of one person or of a small group, but innovation normally requires a larger team, and certainly involves a range of functions and activities. Machiavelli's view of the process is still valid, for human nature does not change, although methods of dealing with opponents are perhaps more humane than was the case in his day.

Important as they are, the human aspects of innovation are largely outside the scope of this book, which is concerned with the financial aspects of innovation. Except in the most trivial cases, financial considerations play a large part in the process of innovation, which normally involves investment in the hope and expectation of an eventual return. The closely-related elements of risk and time are major considerations, and these will continually recur in what follows.

THE INVESTMENT BACKGROUND TO INNOVATION

Innovation can be considered as a special form of investment, and indeed has often been subjected to the types of analysis, some of them very sophisticated, which have been developed for financial investments. The results have not always been happy but, however appropriate these techniques may or may not be in the innovation context, it is necessary to understand them in order to appreciate both their influence and their applicability.

The time value of money

One of the most important ideas which links risk, money, and time is that of the time value of money. This in its turn leads to the concept of discounted ca h flow (DCF), w ich attempts to reduce a stream of more-or-less risky future cash flows to an equivalent current value, known as the present value (PV). If an initial investment I is needed to produce the future cash flow, then the net present value (NPV) is simply:

$$NPV = PV - I$$

The underlying idea is that a certain sum of money available now could be invested or lent to produce an income at an interest rate of r, normally expressed as per cent per year. Thus:

$$\text{Value now (time zero): } = I_0$$

$$\text{Value at the end of 1 year: } I_1 = I_0(1+r_1)$$

$$\text{Value at the end of 2 years: } I_2 = I_0(1+r_1)(1+r_2)$$

and so on, where r_1 is the interest rate in the first year, r_2 is the interest rate in the second year, and so on.

Or, looking at the investment from the other end:

$$\text{Value at the end of 1 year: } = I_1$$

$$\text{Present value (at time zero): } = I_1/(1+r_1)$$

When used to calculate present values from future values, the interest rate is known as the discount rate.

Risk and the discount rate

Interest rates and discount rates are dependent on risk: if there is no practical risk of default on a loan, as with lending to governments of stable countries, then a lender will accept a lower rate than from, say, a company which might get into financial difficulties; similarly, the present value of a sum of money expected at some future date is affected by the apparent likelihood that it will in fact be received. The greater the risk of not getting the money, the more heavily it will be discounted to a present value.

There is thus no single interest rate or discount rate: the appropriate rate varies with the circumstances of the borrower. Interest rates are also affected by the general state of the economy, with the principal influences being the balance of demand for and supply of money and the inflation rate. It is not a trivial matter to decide on the appropriate discount rate to use when using DCF techniques for assessing investments, and we shall have more to say on this subject below. Typical values are in the range 5 to 20 per cent, so that for an investment with a time horizon of several years the effects of the decision can be considerable:

$$\text{For } r = 0.05: \qquad I_0 = 0.78 \ I_5$$

$$\text{For } r = 0.20: \qquad I_0 = 0.40 \ I_5$$

The influence of inflation

Inflation – a rise in the general level of prices, or a decline in the value of money – has a major influence on company operation in general and on discount rates in particular. If a sum of money to be received at some future time is going to buy less than the same amount of money when lent or invested, then the lender or investor will require compensation for the diminution of value as well as for the deferment of the payment. In theory, the adjustment is straightforward – the interest or discount rate is increased by the rate of inflation. In practice, since the interest or discount rate has to be set in advance, whereas the inflation rate is not known until after the event, a new source of risk enters. The problem as far as setting a discount rate is concerned is not so much the inflation itself, but rather its variability. The UK has suffered from comparatively high and variable inflation for some fifty years, with a peak of around 28 per cent and a minimum of zero. It has usually had the highest or second highest inflation rate of the leading industrial countries (HMG 1995).

Uncertainty about inflation tends to push up the discount rate which companies apply to investment appraisal, and thus to make them favour investments which pay off in the shorter term over those which take longer to mature. Another factor which pushes them in the same direction is that inflation affects cash flow, reducing the amount of cash which a company has at its disposal and thus limiting its freedom of manoeuvre. This effect can be simply demonstrated as follows:

Consider a company which makes an operating profit (before tax and interest) of P_0 under conditions of zero inflation, and has borrowings of L on which it pays interest at r_0. If the inflation rate rises from zero to r_i, then:

$$\text{Operating profit: } P_i = (1+r_i)P_0$$

$$\text{Interest paid: } I_i = (r_0+r_i)L$$

The proportion of operating profit paid in interest becomes:

$$I_i/P_i = (r_0+r_i)/(1+r_i)$$

So that:

$r_0 = 0.05$	$r_i = 0.00$	$I_i/P_i = 5$ per cent
$r_0 = 0.05$	$r_i = 0.05$	$I_i/P_i = 9$ per cent
$r_0 = 0.05$	$r_i = 0.10$	$I_i/P_i = 13.5$ per cent

In practice, the effect would be even worse, because the interest rate would almost certainly rise before prices could be raised to restore profit. Not only is the new state after the adjustment to an increase in inflation less favourable, because the proportional increase in interest payable is higher than the proportional increase in the profit from which it is paid, but the transition period is also difficult because the increase in the interest has to be paid before the income to pay it is received.

A combination of inflation and history has resulted in the major companies in the UK being less dependent on long-term loans than is the case in some European countries and Japan. They are instead much more dependent on funds raised from shareholders, whose influence we discuss below. The differences in financial structure between companies in different countries is a significant influence on their behaviour, not least in relation to innovation, and has to be taken into account in making comparisons between patterns of expenditure.

The timing discrepancy also occurs when the national currency is devalued, which is often a result of inflation. Prices of imported raw materials, components, or other supplies rise immediately after a devaluation, whereas any increase in sales arising from the lower prices of exports take time to come through.

These dynamic effects in business are often given less emphasis in textbooks and theories than their importance deserves. Much of economic theory, for example, is concerned with calculating equilibrium conditions of systems which, in practice, are always in transition from one non-equilibrium state to another.

THE COMPANY CONTEXT

Commercial innovations are mainly introduced by firms which are organised as companies. The corporate form offers many advantages, but it also imposes some restrictions. These restrictions affect behaviour, not least in connection with innovative activity, which has no real place in the model of a company which implicitly underlies the laws under which it operates.

The implicit model of a company is of an entity established for a specific purpose in an unchanging world, using capital provided by its shareholders. This capital is used by the company partly for purchasing long-lived tangible assets such as machinery, and partly as the working capital needed to finance its day-to-day operations. Companies are normally 'limited' which means that, in the event of failure, the most that an individual shareholder can lose is the amount subscribed. Companies can either be private, carrying the designation 'Limited' or 'Ltd' as part of their name, or public, carrying the designation 'plc'. Public companies can raise money by offering their shares to the public through the stock market, and those shares may thereafter be bought and sold freely. (There is a technical distinction between stocks and shares which is of no significance for the purposes of this book, in which the terms will be used interchangeably.) Private companies are more restricted in their behaviour.

In return for bearing their share of the risks to which the enterprise is exposed, the shareholders are entitled to vote on major decisions affecting the company, of which the most important in normal circumstances is the election of the directors to run the company; and to receive the profit which it makes. The profit can be distributed as dividend, or it can be retained in the business to finance further expansion. Most often, some of the profit is distributed, adding to the shareholders' income, and some is retained, adding to their wealth.

Definition of profit

The economic definition of the profit for a period is the maximum amount which can be taken out of the company at the end of the period while leaving the company as well off as it was at the beginning of the period. In practice, profit is defined in company accounts in two ways: as the difference between the wealth of the company at the beginning and at the end of the period; and as the excess of the revenue accruing from the activities during the period over the expenses incurred in generating it. The first approach, which was favoured in the early days of company organisation in the mid-nineteenth century, is expressed through the balance sheet. The second approach, which is the primary one today, is expressed in the profit and loss account. Subject to complications which we discuss in Chapter 2 below, the two approaches are equivalent in terms of the figure for profit which they produce, but they emphasise different aspects of company operation. The balance sheet provides a snapshot view of the company at one point in time, while the profit and loss account reveals something about the route by which it got from one point to the next.

Innovation and profit

Other than in the very simplest of transactions, calculating a profit involves some elements of judgement. A whole series of accounting rules and precedents has been generated over the years, the effects of which will be discussed in more detail below, but there is one general rule which has a profound effect on the financial aspects of innovation. This is the rule that states that, before calculating its profit for a period, a company must make provision for replacing its machines and other long-lived assets when they reach the end of their useful lives, but that anything which it spends on replacing its product range, improving the skills and knowledge of its employees, or otherwise preparing itself for a different future, must come off the profit.

This treatment reflects an earlier era in which the tangible assets of a company were its principal sources of earning power. This is no longer true: the assets of a company which determine its earning power are increasingly the intellectual ones such as knowledge, information, skills, organisation, and management capability. With minor exceptions, the conventional accounting treatment of companies acknowledges neither the existence of these assets nor the need to maintain them in order to sustain the company's earning power.

As will be demonstrated in Chapters 2 and 3 below, the prospects are better for repairing the second deficiency than the first.

The importance of intellectual assets has been emphasised by John Kay, one of the UK's leading business economists, in his framework for analysis of the foundations of company success (Kay 1993), which he finds to be unique to each company, but arising from its relationships with its employees, customers, investors, and shareholders. Kay classifies the capabilities which distinguish each company from its competitors into four groups: architecture, which is its set of relationships; reputation; innovation, by which he means the capacity to make innovation as defined by the Innovation Unit; and strategic assets, such as access to mineral or patent rights or control of standards.

In practice, since the world in which the company operates does change, it must make provision for introducing new products or services (which, to save words, we shall hereafter refer to simply as products), and it must update itself in other ways in order to survive, sacrificing some immediate profit in order to do so. More subtly, if it is to survive in an uncertain world, it needs to have some reserve capacity to respond to unforeseen changes. A perfectly-optimised organisation has no capacity to change: it is like a rigid tree which breaks before the storm, whereas a more flexible tree bends to the wind and survives.

Innovation, and the capacity to innovate further, are thus essential if a company is to survive. Since survival is a powerful motive in most companies, they do in practice sacrifice some current profit for the sake of the future, but the balancing act which is necessary to bring the present and the future into some sort of optimum relationship is not easy to perform, and boards of directors have little to guide them except experience and the example of others.

The company and its shareholders

Boards of directors also have to take into account the views of their shareholders who elect them and, more significantly, can sell their shares if the company is public. In return for bearing the risks of the company, the shareholders, as its members, retain ultimate control over the destiny of the company. Legally, the shareholders hire the directors to run the company for them, although in practice it may well be, especially with a small start-up company, that the directors have sought out the shareholders to supply them with capital to expand the business.

The practical significance of the powers of shareholders varies with the circumstances of the company. In small companies there is often a fairly close correlation between the membership of the board and the shareholders. The ability of the shareholders to dispose of their shares is generally restricted, either by the rules of the company or by lack of potential buyers. The shareholders thus have a long-term interest in the company, and will consider the balance between immediate and future profit accordingly. Whether or not the company can in practice survive is another matter, depending mainly on the skills of the management.

The situation is different in large companies which, with few exceptions, are public limited companies whose shares are traded freely on the stock market. Any shareholder who is dissatisfied with the way the company is conducted will find it easier to sell the shares than to change the way the company is being run. This ability to dispose readily of the shares has very profound effects on the company.

The price of the company's shares at any time is determined by the perception of the market about the size and riskiness of the return which they will produce as investments. The return may be in the form of dividends or in a rise in the price of the shares, or in a combination of the two: other things being equal, the division will depend on the dividend policy of the company. A rise in the price of the shares will not just reflect the value of any retained profits, but also any enhanced future earning power arising from new products under development, improvements in efficiency, and many other factors, provided that the market is informed about them.

The information which is available to the market is a critical factor: the company's share price at any time encapsulates the market's view of the future of the company and it will therefore be affected by any change which affects that view. Unexpectedly low or high profits for a year or half-year, announcements about new products, changes in management, or any other change in the company's circumstances or prospects which was not already known will have an effect on share price. Equally, a change which was expected and signalled in advance will normally have no effect on share price.

The nature of shareholders and the market for corporate control

When company law was first enacted in the mid-nineteenth century, most shareholders were wealthy individuals whom the framers of the law envisaged as having a substantial part of their wealth effectively locked into the company. Shareholders were therefore seen as likely to be vigilant in the exercise of their supervisory powers. Under the influence of economic and legislative developments, the situation has changed considerably, particularly in the UK, but also in the US and increasingly in other countries: shareholders are now predominantly institutions of some kind which are investing as intermediaries on behalf of third party beneficiaries to whom they have strict fiduciary duties. Trustees of pension funds, managers of insurance companies or of unit trusts, and others in similar positions have to think first of the interests of those whose money they are investing, not those of the companies in which they make those investments.

In order to safeguard the interests of the ultimate beneficiaries by spreading the risk, professional investors tend to take stakes which are fairly small in relation to the size of the company, in a fairly large number of companies.

With some exceptions, they treat their shareholdings purely as investments and take little part in what has become known as the governance of companies. If a company in which an institution invests gives a good return, then it will keep the company's shares in its portfolio; if the return becomes unsatisfactory, then it will be much more inclined to sell the shares than attempt to exercise its rights as a shareholder to change the management of the company. If the decline in performance has depressed the share price to a point at which the company attracts a bid from another company to acquire those shares, necessarily at a higher price than the current market price, then the institution may well be bound by its duty to its beneficiaries to sell.

There is, in consequence, an active market, not just for shares, but for corporate control. The classical market, dealing in parcels of shares in any one company which are small compared with the total number of shares in issue, has been joined by a market for sufficiently large parcels of shares to give the acquirer the ability to control the acquired company by exercise of its voting rights. Since, if a company is taken over by another, the first casualties are usually the directors, they have a lively interest in maintaining the share price at a level which appears to the market to be as high as is likely to be achieved for the particular business or businesses in which the company operates. This value includes the value of the intellectual assets and bears no necessary or direct relationship to the values shown in the balance sheet.

There are currently some signs that the corporate governance situation is beginning to change in both the UK and the US. In the US, institutional shareholders are obliged by law to exercise their voting rights, and those in the UK are making increasing use of theirs, as well as exercising influence privately by direct contact with boards of directors. As institutional investors have increased their holdings to the point that they own around two thirds of shares of UK quoted companies, they have become increasingly conscious that they cannot rely on finding replacements for those that they regard as unsatisfactory, and that improving the performance of the companies in which they are currently invested is in their interest.

CONSEQUENCES OF THE PROFESSIONALISATION OF INVESTORS

Share prices

A company's share price reflects the market's view of the company's financial prospects. In reaching a collective view, the market is influenced partly by the past performance of the company as expressed in its latest profit figure. Since the profit has to be divided between the shareholders, whose numbers vary very greatly from company to company, investors can compare the past performance of companies on the basis of profit per share. This figure, invariably known as earnings per share (EPS) in deference to US terminology has, until recently, been regarded as a key performance measure. UK companies listed on a recognised stock exchange have been obliged to include a value for EPS in accounts for periods beginning on or after 1 January 1972 under the relevant accounting standard (SSAP 3 1992). The market's view of the sustainability of the EPS is expressed by the multiple of that value which it is willing to pay for a share. This ratio is known as the price/earnings (P/E) ratio, and is usually given in tables of share prices. Clearly:

$$\text{Share price} = \text{EPS} \times \text{P/E}$$

The professionalisation of investors over the last few decades has led to the development of theoretical guidance and valuation techniques. The major concepts are that of the 'efficient market', the valuation of shares on a discounted future earnings basis, and the optimum portfolio.

The efficient market hypothesis

The efficient market hypothesis explains a good deal of the behaviour of stock markets. It is a major influence on the thinking of investors, and hence on the companies in whose shares they invest. Investors do not have to put their money into shares: they can buy real property, such as office blocks and shopping centres, or they can buy government securities (known as 'gilt-edged'), or fixed interest bonds issued by companies or by local authorities. Their choice will be determined mainly by the returns which they expect and the risks which they are willing to take.

An efficient market is defined as one in which information is widely and freely available to all investors and potential investors, who take that information fully into account in arriving at the price at which they are willing to buy or sell. It is important to recognise straight away that the resulting valuation of the shares is not 'correct' in any absolute sense, but is merely the best estimate which the market collectively is capable of making. It is the price at which the demand and supply come into balance at any one time. New information will change the value as will new perceptions of existing information. Stock markets are affected by fashion and sentiment as well as by objective facts about company performance: for example, shares in small biotechnology companies or small computer companies often sell for prices which reflect the perceived promise of future products. Since perception plays a larger part than performance in the pricing of these stocks, they tend to be volatile. At the other end of the scale, stocks in utilities such as gas, water, and electricity companies normally tend to be fairly stable, since their businesses are equally stable. The share prices of privatised UK utilities have, however, been severely affected on occasion by threats of changes to the regulatory regimes under which they operate.

Three forms of the efficient market hypothesis have been distinguished: the weak, semi-strong, and strong forms. The first two forms are well supported by the evidence, but there is some evidence of minor discrepancies between the predictions of the third form and the established facts.

The weak form of the hypothesis states that all available information about the past is fully incorporated into prices, while the semi-strong form says that, additionally, all generally available information about the present is similarly included. The strong form maintains that all information known to anyone in the market, whether generally available or not, is taken into account. It is important to realise that 'available' in the stock market context means available to the market, through which information travels extremely rapidly: it does not mean available to the general public, whose reliance on the news media means that new information will usually have affected share prices before they become aware of it. It is the rapid response of the market which makes it efficient.

The weak form of the hypothesis implies that any attempt to beat the market by predicting share prices on the basis of past behaviour will fail. All the available evidence is that price changes follow a random walk pattern, in which any change is independent of the previous change. (It must, of course, start from where the previous change finished.) Information about past prices is therefore useless in predicting the future. Some very minor exceptions, such as an abnormal rise in the shares in small companies in the first few days of January have been detected but these are not usually large enough to give a return which exceeds the dealing costs involved in taking advantage of them. In any case, should the existence of such an effect become generally accepted, then it would be eliminated by the normal workings of the market: the pressure of traders attempting to buy shares in anticipation of a price rise would trigger off that rise.

The semi-strong form of the hypothesis implies that any attempt to beat the market by careful study of company accounts and other available current information will be similarly unsuccessful. Studies of the empirical evidence again seem to support this form of the hypothesis although there may be a somewhat slower adjustment to some forms of new information than others. Again, the effects are not large enough to offer opportunities for profitable trading.

The strong form of the hypothesis implies that information which is not generally known will be impounded in the share price as soon as it is used as the basis of dealing. It is not necessary for those in the market who are not in possession of the information to know what that information is: the actions of those who are will affect the market in a perceptible way.

There is thus an opportunity for those in possession of inside information – such as of an impending takeover bid, or of unexpectedly high impending profits – to profit from that information, but the opportunity will be denied to others. It is for this reason that insider dealing has been prohibited by law. Cases of anticipation of announcements do, however, still occur, possibly stimulated by rumour rather than dealing.

Professional investors who take a good deal of trouble to acquire extra information as the basis of their investments do seem to obtain slightly better returns than a randomly-selected portfolio of stocks would have given them, but not to a sufficient degree to outweigh the cost of acquiring the information, and seldom on a consistent basis from year to year. The strong form of the efficient market hypothesis thus receives a good deal of support. Although it is clearly not in the interests of professional investors to accept it, some of them have bowed in its direction to the extent of setting up 'index linked' funds whose composition mirrors that of some stock market index, and must then move in step with it. Although not being able to hold out the hope of beating the index, such passive investment funds can at least promise to keep pace with it.

Share prices are thus essentially determined by the general perception of the market: an investor aiming to make a profit by buying a share which that investor's analysis considers to be undervalued has to wait until the perceptions of other investors come into line before the profit can be realised.

The general perception of the market can change quickly on occasion. One such was the 1987 stock market crash which started on Wall Street in New York and rapidly spread round the world. The crash was attributed by some to the effects of traders following formulas which set trigger points for selling, and which were implemented automatically. As these conditions did not hold in all the markets which experienced the crash, this explanation is not convincing. A plausible case has been made that the real trigger was the belated realisation that a rise in interest rates and an increase in the general level of risk affecting companies in an increasingly uncertain world genuinely reduced the value of shares.

Assuming that the price of a share at any one time is the discounted value of the future stream of dividends (an issue which will be explored in more detail below), we can express the current share price S_0 as:

$$S_0 = D_0[(1+g)/(1+r) + (1+g)^2/(1+r)^2 + (1+g)^3/(1+r)^3 + ...]$$

where: D_0 = the current dividend

r = the discount rate

g = the expected growth rate of the dividend

This series sums to:

$$S_0 = D_0(1+g)/(r-g)$$

The growth rate g and the discount rate r will both typically be in the fairly narrow range of 0.05 to 0.25. Their difference, $r-g$, which determines S_0, will therefore be sensitively dependent on changes in either value. For example, if the expected growth rate before some change was 5 per cent and the discount rate was 15 per cent, whereas after it the growth rate was reduced to 4.5 per cent and the discount rate increased to 18 per cent, the relative change in the share price would be:

$$S_a/S_b = (1+0.045)(0.15 - 0.05)/[(1+0.05)(0.18 - 0.045)]$$

$$= 0.74$$

This drop of 26 per cent is about that which occurred in the 1987 crash.

Seen in this light, the sensitivity of share prices to comparatively minor items of news becomes much more understandable.

Investors thus need to look at the stock market as a whole, and at the stock market as one element in an investment portfolio. Theories have been developed for structuring and valuing portfolios of stocks, a subject to which we now turn.

Portfolio construction and valuation

The valuation of financial assets such as shares, government securities, and bonds is made easier by the very large amount of historical data which is available. The returns and the variability of those returns can be readily calculated by standard statistical techniques. As would be expected, the return and risk increase together. Treasury stock gives returns which usually do little more than compensate for inflation, but do it fairly reliably; government bonds and company bonds give real (after inflation) returns which are around 2–3 per cent, with rather more variability from time to time, whereas ordinary shares give higher returns, on average perhaps 8–10 per cent higher than the risk-free base line of Treasury stock, but at the price of considerable variability.

Individual share prices are affected by two sets of factors: those which affect all shares, and those which are specific to the share in question. The first set gives rise to what is known as market risk, while the second set produces the specific or unique risk peculiar to that share.

A change in the interest rate set by the government, central bank, or other national authority will affect all share prices because it upsets the established risk premium. If the rate goes down, then share prices will rise to reduce their return and restore the differential; if the interest rate goes up, share prices will drop, again to preserve the relative return. Other changes, such as the announcement of a general election, which introduces an element of uncertainty as to its outcome, will also affect all share prices.

Changes in the circumstances or prospects of an individual company will, however, primarily affect the price of that company's shares. There may be some effect on the share prices of companies in the same sector of business if the change is one of general application to that sector, but generally share price changes in companies have a unique element. On any one day, or over any particular period, movements of individual share prices will have an element which is uncorrelated. As one share is going up, another will be coming down.

The next day, the situation may be reversed. Because of this spread, the variability of the market as whole, despite being subject to a general market risk, will be much less than the variability of most of the individual shares of which it is composed.

This effect can be exploited by individual investors who, although unable in practice to buy shares in every quoted company, can reduce the overall risk of their individual portfolios by buying a range of shares. In practice, a diversified portfolio of something like 15–30 shares will reduce the variability of the portfolio to the practicable minimum.

An investor has a further factor to consider. This is the general level of risk to which the portfolio is subject. Diversification removes individual or unique risk but, by definition, leaves the portfolio vulnerable to market risk. Although share prices are equally subject to market risk, the degree to which they are subject varies. Some shares are much more sensitive than others to general changes affecting the market as a whole. It is thus possible to construct diversified portfolios with different degrees of overall risk.

The ratio of the sensitivity of the individual share price to changes in the general level of market prices is known as its beta (β). The average stock thus has a beta of 1, while a sensitive stock will have a beta of more than 1, and an insensitive stock a beta of less than 1. It is intuitively clear, and can be proved rigorously, that the risk of a diversified portfolio is proportional to the average beta of the stocks of which it is composed.

Investors accept risk in the hope of a return and the higher the risk the higher the return they expect. They therefore seek to acquire a portfolio which gives the best combination of risk and expected return for their purposes: investors have different priorities, which is just as well for the existence of an active market. Given the existence of a range of stocks with different combinations of risk and expected return, it is possible to construct a very large number of different diversified portfolios with more attractive combinations than that of any individual stock, but there is a limit to how much improvement is possible. Portfolios with the best expected return for a given risk are known as efficient portfolios.

The capital asset pricing model (CAPM)

It is possible to extend this analysis further by taking into account that investors do not have to put all their money into the stock market. If they do not want to expose all their money to even the minimum risk which a diversified portfolio of low-risk stock offers, they can invest some of it and lend the rest to get a guaranteed interest income. Alternatively, if they are willing to accept even higher risks but do not find the expected returns from the corresponding diversified portfolio high enough, they can borrow money at fixed interest to invest more in the stock market.

As mentioned above, there is good statistical evidence which shows that the market expects a return on common stocks which is around 8–10 per cent above the return available on risk-free investments. This market risk premium applies to the market as a whole, not to any specific stock. The relation between the risk premium for the market as a whole and that for an individual stock is given by the capital asset pricing model (CAPM), a very significant influence on investors' thinking. The model says, very simply, that the risk premium for a stock with a given beta is β times the risk premium for the market as a whole. This result follows from the argument that in the absence of better information than is available to the market, the market portfolio gives the best combination of risk and expected return. Since the contribution to the variability of the market portfolio arising from any particular stock is proportional to its beta, the risk premium demanded on it should also be beta. If it were not so, then it would either be so good an investment that the market would soon adjust the price, or so bad an investment that the price would drop.

The CAPM seems to be reasonably well-supported by empirical evidence. As it deals with expected returns it is not strictly testable, but it seems that beta figures derived from past performance are a good guide to the return in the subsequent period. The fundamental argument that investors expect a higher return on riskier stocks is hardly contestable, and the empirical evidence supports the view that they are more concerned with market risk than unique risk. They can diversify their investments themselves and do not put any higher value on a diversified company than on the sum of its separate components.

There are other theories of market pricing which the interested reader can study in works on financial theory such as Brealey and Myers (1991). The purpose of the above short account is to acquaint the reader with the basic concepts which underlie investment and hence affect the way in which investors perceive companies.

The SVA and EVA approaches

Two recent developments in assessing the value of companies place emphasis on the sustainability of earnings and hence take into account the innovative activities which will sustain those earnings into the future.

Shareholder value analysis (SVA) is based explicitly on the view that the value of a company to an investor consists of the discounted value of the stream of future dividends plus the capital gains arising from share price increases, themselves reflecting future prospects.

The economic value added (EVA) approach recognises that modern shareholders demand a fairly steady return on their investment, and are not content to receive a fluctuating, and possibly non-existent, annual reward in the form of a share of the profits, as the classical model of a company envisages. The company must instead recognise that it has to pay the going rate for its funds, whether borrowed at bank interest or invested by shareholders. The true surplus, or economic value added by the company, is what remains when all expenses have been met, including the costs of the capital used by the business. On this view, dividends become a more-or-less fixed expense of running the business.

Kay (1993), who essentially adopts the EVA approach, rather confusingly uses the term 'added value' for the same concept. This concept is not the same as that of 'value added', which is discussed further in Chapter 2.

In seeking to value companies on either the SVA or EVA bases, which are not incompatible, analysts place heavy emphasis on the cash-generating capacity of the company since it is only from cash that dividends can be paid and investments made in order to sustain those dividends into the future.

SHORT-TERMISM

Most of the account of the relationship between investors and companies given above is generally agreed, but a controversy rages about whether investors in practice demand returns which are too high for the ultimate long-term health of the companies in which they invest. This argument over 'short-termism', as it has inelegantly come to be known, is often rather sterile (Marsh 1990) but, as innovation is one of the key issues, and there is evidence that innovation behaviour is affected by beliefs that short-termism exists, some account of it is relevant here.

Professional investors are interested in the returns that will be obtained from their investments: they are not interested in the companies in which they invest, except as sources of returns. Their primary responsibility is to those whose money they are investing, not to the companies in which they invest. Indeed, given that for the portfolio planning reasons explored above they diversify their investments and seldom own more than a very small proportion of the shares of any one company, the amount of time and energy which they can devote to their responsibilities as members of those companies is necessarily limited. They are under strong pressure to get the best return that they can, and under constant threat of having funds removed from their care if their investment performance fails to meet the expectations of those for whom they are acting.

It is widely believed that the result is that professional investors demand too high an immediate return from companies, forcing them to maintain the dividend, if necessary at the expense of making provision, especially by way of expenditure on innovation, for the future. This phenomenon, known as 'short-termism' has been widely discussed as a shortcoming of the Anglo-Saxon (UK and US) system of financing companies, and contrasted unfavourably with the perceived 'long-termism' of investors elsewhere, especially in Japan and Germany.

The issue raises strong emotions on the parts of both industrialists and the investment community, and there has been a good deal more heat than light in much of the discussion. Dispassionate studies indicate that short-term pressures undoubtedly exist inside companies, but the evidence is weak that these pressures actually emanate from investors. It seems that substantially more of the problem, and even more of the opportunity for reducing it, lies within companies than outside them.

The key to resolving the problem seems to be to improve mutual understanding and communication, an end to which it is hoped this book will make a contribution.

Managers' views

Strong evidence of the existence of short-term pressures within companies emerged from a postal survey of the finance directors of the (UK) Times 1000 companies (Collison, Grinyer, and Russell 1993). The survey also showed that the source of these pressures was the directors' perceptions of what the attitudes of investors (collectively known as 'the City') would be. From the 246 replies to a series of questions, some particularly relevant findings were:

- 66 per cent believed that the capital market valued a company primarily by reference to its prospective earnings for the current year.
- 67 per cent agreed that undertaking an innovative project which resulted in a 15 per cent drop in the net earnings expected by the capital market would adversely affect the company's share price.
- 53 per cent agreed that top management would not accept proposals for increasing expenditure on R and D if it resulted in a significant fall in earnings growth below capital market expectations.
- 50 per cent said that management would require a major new product to start to generate accounting profits within two years, and a further 32 per cent said that the figure would be three years.

A significant feature of the findings of this survey is the emphasis on investors' expectations, as perceived by managers. Other surveys have shown that managers tend to have few contacts with investors whom they tend to regard as a distant and alien tribe. Such surveys also find that managers regard cutting costs as one of their highest priorities and that they regard research and development as one of these costs rather than as a form of investment.

Investors' views

The investors' view was tested in a study of the data for the stock market prices of 477 non-financial UK companies, together accounting for about 50 per cent of the market value of all quoted UK companies, for the period 1980–1988 (Miles 1993). On the assumption that the price at any one time represented the discounted value of the future earnings, and allowing for tax and risk factors by using a modified version of the CAPM equations for price were set up on various assumptions and tested for fit with the data. It was found that the best fit was for the assumption that the stock market applies a discount factor of around 15 per cent for the first five years' future earnings, and a discount rate of around twice as much – 30 per cent – for the next five years' earnings. Earnings beyond ten years at a discount rate of 30 per cent or more are, of course, of negligible current value.

The conclusion from this study was that while the evidence supported the view that longer-term cash flows are heavily discounted by the market, it did not follow that the market was wrong to do so. Even so, the evidence was that the City takes a longer-term view than the industrial respondents quoted above believed it to do.

In practice, since all companies are treated equally, an extra discount on earnings beyond five years can have an adverse effect on the valuation only of those companies who are depressing current profits by making significant revenue investments (such as R and D) in the expectation of making an enhanced profit after five years.

The case for the existence of short-termism is thus established. The evidence that it results from real pressures on companies from shareholders, as opposed to perceived pressures is, however, weak. The reluctance of shareholders to attach much value to company earnings beyond five years may well reflect not lack of confidence in the companies themselves but justified fears about inflation, changes in government policy, and other general economic factors.

Evidence for long-termism

There is, however, evidence in the other direction. Opinion surveys of company directors and investors show that, if anything, the investors put a higher priority on innovation and on research and development than do the industrialists (BQMA 1993, 1994). Some companies do make long-term investments and are supported by the City in doing so. The pharmaceutical industry is the best-known case. It is undoubtedly an industry in which returns take a long time to accrue, and in which heavy expenditure on R and D and other aspects of innovation – especially marketing – are required, but it is also one in which some UK companies have been successful on a global basis.

The pharmaceutical industry does not stand as the sole example. Several studies of engineering and other non-pharmaceutical companies which have prospered in the UK financial climate, sometimes improving their performance very considerably in the process, have shown that long-term strategies played a major part in their success. Their boards had assessed their companies' long-term business needs in the light of world competition and taken appropriate action, sometimes depressing immediate profits in order to do so (Fellowship of Engineering 1991). None of these companies reported any complaints about the attitudes of their investors. They did, however, take care to communicate their strategy and their progress in implementing it to their investors, keeping them fully in the picture throughout.

The evidence thus seems to be that some UK companies, perhaps the majority, see the City as imposing pressures for short-term performance, while a minority of successful companies do not. The main difference between the two groups seems to be that the successful and innovative companies communicate with their investors while the less successful do not. The successful and innovative establish mutual understanding with their investors while the less successful have relations which are much less close and supportive.

COMMUNICATION BETWEEN COMPANIES AND SHAREHOLDERS

Communication is the key to reducing the effects of short-termism. It would be too much to hope that complete harmony between companies and investors could be achieved in all cases and over all time, particularly as there are some divergences of interest over the relative importance of dividends and re-investment.

A substantial proportion of UK investors, such as pension funds, enjoy tax concessions which distort the theoretical tax equality between profits distributed as dividends and profits retained in the business for expansion. When a company pays a dividend it also pays another 25 per cent in advance corporation tax (ACT). As its name suggests, the company can count ACT against the tax which it eventually becomes liable to pay on profits. The recipients of the dividend can count the ACT as part of their tax payment but, if they are not liable to tax, can reclaim it from the government. In effect, they pay less tax, and thus obtain a better return, from profits paid out to them as dividends than from the same profits retained. Even without this distortion they may also believe that profits returned to them in dividends would be better invested elsewhere than retained in the business, either because the business is mature, with few opportunities for profitable expansion, or because they have less confidence than the management in the management's ability to use the retentions wisely.

The last point illustrates the importance of improving the mutual understanding between shareholders and the companies in which they invest. The investor's role in a company is essentially a passive one: apart from occasional exercise of the shareholder's right to vote, usually on matters of purely technical significance, the investor can exercise no control over the workings of the company, and must therefore look upon it primarily as a source of income. Discounting the value of that income at the relevant rate and comparing the return with what could be obtained elsewhere is the appropriate technique to use. If the investor does not like the result, the best action is to sell the shares and invest elsewhere. Investors' activities are about moving money from one investment to another with monitoring of what happens.

Managers, on the other hand, have an active role: they can control what happens. If they do not like the predictions of a discounted cash flow analysis of the business, or of a specific project within it, they have the power to change the business or the project to produce a better outcome.

Expenditure on innovation must be a key element of the dialogue between companies and their shareholders, together with long-term capital investments. Expenditure on both is to some extent made at the expense of the present, and getting the balance right is a matter of judgement. If interest rates are high, or the general uncertainty in the economy is great then, on a purely discounted cash flow basis, no long-term investment is worth making. On the other hand, present products will not last indefinitely into the future as overseas competitors will not stop introducing new products, and the company will have no future without new products. The implicit assumption that the current business will continue indefinitely, and that new investment can therefore be considered in isolation, must be recognised as false by all concerned.

A concept which has the potential to unify the points of view of all those with the long-term interests of the company as their objective is that of maximising its net present value, discounted at its cost of capital.

Accounting provides much of the language in which dialogue between shareholders and companies is conducted. We therefore now turn to a more detailed look at some accounting issues, starting with the financial accounting through which a company reports to its shareholders and creditors, and then looking at basic techniques of valuation which form a bridge between these accounts and our final accounting topic, the management accounting which a company uses for its internal affairs.

Questions for discussion

1 Sound recording was originally achieved on wax or tinfoil cylinders. Successive generations were shellac discs, vinyl discs, magnetic tape, and compact discs. How could these changes be classified in terms of the fundamental, radical, and incremental degrees of innovation? Where does digital audio tape fit?
2 Does the 'short-termism' dilemma affect only a business which is undergoing a step change?
3 Would it be desirable to restore effective tax neutrality between dividends and retentions? If so, how?

Chapter 2

Innovation and financial accounting

ACCOUNTS AND ACCOUNTABILITY

Accounts are records of financial transactions. They are essential for the proper management of any business and as a medium for reporting on the conduct and results of that business. In the first application they are known as management accounts and are, as the name suggests, of primary interest and value to the managers of the business who have a good deal of freedom to decide on their scope, form, and content. The management accounting issues of concern in the innovation context are considered in Chapter 4.

In the second application – that of reporting to those outside the business who may have a legitimate interest in its progress – the relevant accounts are known as financial accounts, which form the subject of this chapter. As with management accounts, the responsibility for the preparation of financial accounts falls to the managers but in this case their freedom to decide on the scope, form, and content of the accounts is considerably circumscribed by law and regulation. Since the performance of managers is largely judged by the financial accounts which they produce, they tend to be very conscious of the effects of any business decision on the appearance of those accounts. Those doing the judging are very often investors, and particularly the professional investors whose behaviour was discussed in Chapter 1 above. Such investors tend to regard the business very much as it appears from the financial accounts. In extreme cases, the accounts may be seen almost as being the business.

It is, in consequence, difficult to over-estimate the importance of financial accounts. The rules by which they are drawn up are therefore of considerable significance and it is the purpose of this Chapter to examine some of the main issues with special reference to their effects on innovative activity. Since most business activity with a significant innovation content is in the hands of companies, the focus will be on the rules as they affect companies.

The requirement for financial accounts

Those outside the business who have a legitimate right to be informed about its progress are primarily those who provide it with capital, either on a temporary or permanent basis, and it is primarily to them that the financial accounts are directed. The tax authorities also have an interest. The financial health of a company is also of legitimate interest to its customers, its employees, and the local community or communities in which it operates. The importance to the company of these other groups of 'stakeholders' as they are sometimes called is increasingly being recognised, but the response to that recognition has not involved any major changes in accounting practice as yet.

A company is an artificial creation of the law, a 'legal person' which has an existence independent of that of its members or shareholders, who may change individually or even as a body without affecting the existence and identity of the company. The company's constitution is set out in two documents. The Memorandum of Association specifies its name, the site of its registered office (England and Wales, Scotland, or Wales), its objects, whether it is private or public, whether the liability of members is limited, and the amount of share capital. The Articles of Association contain the detailed rules for the conduct of the company and, in particular, the powers of the directors. There is some freedom for companies to vary the Articles to suit their own circumstances, but there is a standard form which will be assumed to apply in what follows. Readers with an interest in the details of company law should consult specialist texts, such as Gower (1992), on this complex and ever-evolving subject.

The shareholders of a company provide the initial capital on which the business is founded. In return they expect to receive income in the form of dividends to be paid out of the profit the company makes. Except in special circumstances, the shareholders cannot get their money back from the company, but they can achieve the same object by selling their shares, together with the entitlement to the dividend, to someone else. If the company is a public one listed on a stock exchange, then selling the shares is straightforward, for there will always be a dealer ready to buy, at a price. If the company is public but unlisted, then finding a willing buyer may require some search and negotiation. If the company is private it cannot offer its shares for sale to the public, the market for them is likely to be very small, and there may well be restrictions in the Articles of Association on their disposal, such as a requirement to obtain the approval of all the other shareholders to the sale.

The management of the business of a company is entrusted to the directors who are appointed by the shareholders and can be removed by them. Despite popular belief, the duty of the directors is not owed solely to the current shareholders, but to the company. The interpretation to be put upon 'the company' depends to some extent upon the circumstances, but there is no doubt that it is the continuing entity, a concept which is important in the context of innovation, as will be discussed below. The directors have to report periodically to the shareholders about their stewardship of the business, not least through the medium of financial accounts, on the basis of which they recommend what dividend should be paid.

The accounts of a company have to be audited by a qualified person appointed, at least in legal form, by the shareholders. The auditor has to certify that the accounts give 'a true and fair view'. This concept, which is legally regarded as being of overriding importance, has defied precise definition, but has the virtue of making it clear that accounts are, to some extent at least, a matter of opinion. The law does not require the auditor to certify that the accounts give the only possible true and fair view, but merely one, possibly of several or even many views, which satisfies those criteria.

If, as is usual, the company is a limited company in which the liability of the members is limited to the amount they have subscribed, then trade creditors, banks, and others with any claims on the company can have recourse only to the company, not to its members, in the event of default. They therefore have a legitimate interest in its financial health as expressed in its accounts. Since it is impossible to know who such people might be from time to time, a limited company has to deposit a copy of its accounts in a registry to which any interested party can have access. The accounts of a limited company are thus public property.

In practice, the larger, listed, companies have so many shareholders that they have to print their accounts in large numbers, usually as part of an annual report which includes other material, some of which is at the discretion of the directors. Such reports are usually made available to any member of the public on request, and some companies go so far as to invite requests on the occasion of the announcement of their annual results.

The law permits companies to be shareholders in other companies, and this freedom has been widely exploited to form groups of companies in which the operating companies are subsidiaries of the parent or holding company. It is usually this holding company which is listed on the stock exchange. Although the individual companies have to file accounts, it is the consolidated group accounts which attract attention. The subsidiary companies may be partly owned by others than the group parent, giving rise to what are known as minority interests in the group's assets and profits.

To summarise thus far: preparing accounts for shareholders is the price of incorporation, and publication of accounts is the price of limited liability.

The form and content of accounts

Other than a vague requirement that dividends should be paid out of income or profits, and not out of the company's capital, which should be preserved, the law until comparatively recently left decisions on the form and content of company accounts to custom and practice as determined by accountants. The discretion thus left to individual company managements led to considerable variation in practice which became of some significance when takeovers became more common. Successful predators sometimes found – and complained bitterly in consequence – that the businesses they had taken over were not as profitable, by the acquirer's standards, as they had been by those of the previous management, on which they had at least partly relied in making their bid. In response, and in the tradition of self-regulation and minimal legal interference, the accountancy profession in 1970 set up a mechanism for setting accounting standards which would be enforced by professional discipline on the preparers and auditors of accounts. The series of standards produced by the Accounting Standards Steering Committee (ASSC), which later became the Accounting Standards Committee (ASC) are known as Statements of Standard Accounting Practice or SSAPs.

A major change occurred under the Companies Act 1981 which fulfilled an obligation on the UK government to implement the Fourth European Community Directive on Company Law, which concerned itself with the form and content of company accounts. This Act laid down certain fundamental principles of accounting and also specified the form of the balance sheet and the profit and loss account. It also reaffirmed the primacy of the 'true and fair' concept, whose inclusion in the Directive had been secured by UK negotiators.

The 1981 Act, which was subsequently consolidated into the Companies Act 1985, provides for exemption from some of the reporting requirements for small and medium-sized private companies, defined as those which meet two out of the three criteria in Table 2.1.

Table 2.1 Definitions of small and medium-sized companies

Criterion	Small company	Medium-sized company
Turnover	Not more than £2.8M	Not more than £11.2M
Balance sheet total	Not more than £1.4M	Not more than £5.6M
Number of employees	Not more than 50	Not more than 250

Source: Companies Act 1985, as amended

The monetary figures are adjusted from time to time in line with inflation and in accordance with harmonised European requirements, which are expressed in ECU, and consequently give rise to the rather arbitrary-looking values.

The balance sheet total refers to the company's assets, which are items A to D in the standard form of the balance sheet shown in Table 2.2.

A further Directive, this time concerned with the accounts of groups of companies, was enacted in the Companies Act 1989. This Act also gave statutory recognition to accounting standards, in particular by requiring that company accounts should state whether they had been prepared in accordance with applicable accounting standards, and giving details of any material departure from those standards and the reasons for so doing. The Act also introduced a provision whereby the compliance of a given set of accounts with the 'true and fair' view could be tested in the courts. Legal opinion is that these provisions effectively give statutory backing to accounting standards which, since 1990, have been produced by the Accounting Standards Board (ASB), a body which represents not only the accounting profession but also the

government and producers and users of accounts. The ASB's standards are
known as Financial Reporting Standards (FRSs). Both SSAPs and FRSs are
current.

The legal rules controlling companies are expressed partly in the main Act
and partly by regulations made under the Act. The Secretary of State has
express powers to vary the requirements for the content of company accounts
by regulation.

A further influence on companies whose shares are listed on a stock
exchange is exercised by the rules laid down by that exchange. In the case of
the London Stock Exchange, these rules impose some further requirements
on the detail to be included in accounts and an obligation to produce a
half-yearly financial report covering the main items in the financial accounts.
The London Stock Exchange is traditionally reluctant to include provisions
in its Listing Agreement at the behest of outsiders who see it as an alternative
means to government or professional regulation for introducing reporting
requirements – not least of information about various aspects of innovative
activity – which they regard as being desirable.

The financial accounts of companies are thus governed by law, regulation
under the law, accounting standards, and the requirements of the stock
exchange.

FINANCIAL ACCOUNTING

Some basics of financial accounting

This section gives a brief account of the basic concepts which underlie
financial accounting and which need to be understood to appreciate the
influence of accounting on innovation. For full descriptions and explanations,
readers should consult a textbook on accounting, such as Berry (1993).

The basis of all accounting is the concept that a transaction – which by
definition involves two parties – has two aspects, both of which must be
recorded. This principle of duality is the basis of double-entry book-keeping,
first codified in 1494. A simple example would be the purchase of a machine
using cash in the company bank account. The two aspects of this transaction
are the increase in the company's assets by the cost of the machine, and the
diminution in its bank account by the same amount. The recording of all
transactions in this way enables the company to keep track of its financial
relations with the various parties with whom it deals and provides a checking
mechanism to ensure that all transactions have been recorded.

A second important concept is the distinction between capital and revenue.
Capital items, such as machinery and factories, have a value which outlasts
that of the period for which accounts are being drawn up: revenue items, such
as wages and other employment expenses or sales income, are completely
contained within the period. This distinction is reflected in the two basic

financial statements – the balance sheet and the profit and loss account.

The balance sheet records the assets and liabilities of the company at a given point in time, usually the end of its financial year, with the difference between them being the net wealth of the company, also known as the shareholders' funds or owners' equity. This total includes not only the original capital of the company but the accumulated undistributed profits, and the terminology reflects that in the event of a liquidation the remaining funds belong to the shareholders. FRS 5 (1994) defines assets as:

Rights or other access to future economic benefits controlled by an entity as a result of past transactions or events.

It defines liabilities in parallel terms as:

An entity's obligations to transfer economic benefits as a result of past transactions or events.

The profit and loss account lists the company's income and expenditure for a period, usually the financial year, with the difference between them being either a profit or a loss. The principle of duality, expressed through double-entry book-keeping, ensures that the change in the company's wealth from one year end to another, as shown in the respective balance sheets, is the same as the profit or loss shown by the profit and loss account for the period between the two balance sheets.

Four basic concepts or principles of accountancy were defined, on the basis of long usage, in SSAP 2 (1971). These are the principles of the going concern, accruals, consistency, and prudence. Since the 1981 Companies Act these principles have been enshrined in law. Adherence to the first and third principles ensures that the accounts of a company are prepared on the assumption that its business will continue indefinitely and on a consistent basis from year to year. The prudence principle, in the words of the Act means that:

(a) only profits realised at the balance sheet date shall be included in the profit and loss account; and
(b) all liabilities and losses which have arisen or are likely to arise in respect of the financial year to which the accounts relate or a previous financial year shall be taken into account, including those which only became apparent between the balance sheet date and the date on which it is signed on behalf of the board of directors ...

The accruals basis is expressed by the Act as:

All income and charges relating to the financial year to which the accounts relate shall be taken into account, without regard to the date of receipt or payment.

Income and charges are matched as far as possible, so that the charges relating to a particular item of income are accounted for in the same period.

Both this matching principle and the prudence principle are important in the context of innovation, as will be seen later.

The scope for judgement which accounts allow is clear from these definitions.

The balance sheet

The minimum content of the balance sheet, as specified by company law, is shown in Table 2.2. The items of most interest in the context of innovation are those included under the heading 'intangible assets': development costs; patents and related rights, which are usually known as intellectual property rights (IPR); and goodwill.

Company law, in conformity with the requirements of the Fourth Directive – which means that it cannot be changed without European negotiation – forbids the capitalisation of research costs. This treatment is in accordance with the prudence principle, since the future economic benefits arising from research are highly uncertain at best.

As can be seen from Table 2.2, the law allows development costs to be capitalised and included on the balance sheet. The Companies Act provides, however, that this freedom can be exercised only 'in special circumstances'. Even then, under rules designed to ensure that companies do not pay dividends or make other distributions out of capital, the capitalised development expenses are deemed to be realised losses, and hence not distributable. If the directors do decide to capitalise development costs, they must include in the accounts a note giving the period over which the amount originally capitalised is being written off (by charging it to the profit and loss account), and the reasons why it was capitalised in the first place.

The 'special circumstances' in which capitalisation of development costs is permissible are defined by SSAP 13 (1989), which governs the accounting treatment of research and development (R and D) expenditure. SSAP 13 was originally issued in late 1977 and revised in early 1989. It specifies that development costs may be deferred (a term which it prefers to 'capitalised', reflecting the restrictions on distributability) where:

(a) there is a clearly defined project, and
(b) the related expenditure is separately identifiable, and
(c) the outcome of such a project has been assessed with reasonable certainty as to:
  (i) its technical feasibility, and
  (ii) its ultimate commercial viability considered in the light of factors such as likely market conditions (including competing products), public opinion, consumer and environmental legislation, and
(d) the aggregate of the deferred development costs, any further development costs, and related production, selling and administration costs is

Table 2.2 Balance sheet format 1

A   Called up share capital not paid
B   Fixed assets
    I   Intangible assets
       1   Development costs
       2   Concessions, patents, licenses, trade marks and similar rights and assets
       3   Goodwill
       4   Payments on account
    II Tangible assets
       1   Land and buildings
       2   Plant and machinery
       3   Fixtures, fittings, tools, and equipment
       4   Payments on account and assets in course of construction
    III Investments
       1   Shares in group undertakings
       2   Loans to group undertakings
       3   Participating interests
       4   Loans to undertakings in which the company has a participating interest
       5   Other investments other than loans
       6   Other loans
       7   Own shares
C   Current assets
    I   Stocks
       1   Raw materials and consumables
       2   Work in progress
       3   Finished goods and goods for trade
       4   Payments on account
    II Debtors
       1   Trade debtors
       2   Amounts owed by group undertakings
       3   Amounts owed by undertakings in which the company has a participating interest
       4   Other debtors
       5   Called up share capital not paid
       6   Prepayments and accrued income
    III Investments
       1   Shares in group undertakings
       2   Own shares
       3   Other investments
    IV Cash at bank and in hand
D   Prepayments and accrued income
E   Creditors: amounts falling due within one year
       1   Debenture loans
       2   Bank loans and overdrafts
       3   Payments received on account
       4   Trade creditors
       5   Bills of exchange payable
       6   Amounts owed to group undertakings
       7   Amounts owed to undertakings in which the company has a participating interest

      8   Other creditors including taxation and social security
      9   Accruals and deferred income
F   Net current assets (liabilities)
G   Total assets less current liabilities
H   Creditors: amounts falling due after more than one year
      1   Debenture loans
      2   Bank loans and overdrafts
      3   Payments received on account
      4   Trade creditors
      5   Bills of exchange payable
      6   Amounts owed to group undertakings
      7   Amounts owed to undertakings in which the company has a participating interest
      8   Other creditors including taxation and social security
      9   Accruals and deferred income
I   Provisions for liabilities and charges
      1   Pensions and similar obligations
      2   Taxation, including deferred taxation
      3   Other provisions
J   Accruals and deferred income
K   Capital and reserves
    I   Called up share capital
    II Share premium account
    III Revaluation reserve
    IV Other reserves
      1   Capital redemption reserve
      2   Reserve for own shares
      3   Reserves provided for by the articles of association
      4   Other reserves
    V Profit and loss account

[Minority interests]

Source: Companies Act 1985, as amended
Note: There is an alternative format, which contains the same items, but in a different order. In the alternative version the asset items (A to D) are grouped together, followed by the liabilities items (K, I, E+H, and J). Although this separation reflects the pure theory of the balance sheet, the version given above provides figures for net current assets (or liabilities, as the case might be) and for total assets less current liabilities, which are figures useful for analysis. Probably for this reason, it seems to be the more popular amongst UK companies.
The reference to minority interests is required for group accounts.

reasonably expected to be exceeded by future sales or other revenues, and
(e) adequate resources exist, or are reasonably expected to be available, to enable the project to be completed and to provide any consequential increases in working capital.

These conditions are fairly restrictive but they incidentally provide a useful check list of the financial aspects of innovation. The policy issues surrounding capitalisation of development expenditure will be discussed below, but it can be noted now that the practice is rare. For example, the annual survey of financial reporting for 1991–92 (Skerratt and Tonkin 1992) found that of the 100 large listed companies whose accounts it surveyed, sixty-two showed evidence of R and D expenditure, but only five showed evidence of capitalisation. The figures for 100 medium (but still in the Times 1000) listed companies were five out of forty-two, and there were none in the thirty-four out of a sample of fifty large unlisted companies.

The Companies Act permits the capitalisation of patents and similar rights if they have either been acquired for valuable consideration and are not required to be included under goodwill, or they were created by the company itself. The distinguishing feature of such rights is that they are protected by law in some way, and are thus definable fairly precisely, making them relatively easy to buy (for 'valuable consideration') or sell in a way which would not normally be possible with a development project.

Goodwill is a term used for the difference between the purchase price of a business and the value of its assets. It can be thought of as 'intellectual capital', representing the value of the company's knowledge of its customers, the skills and knowledge of its workforce, its systems of operation, and all the other intellectual assets which, as discussed in Chapter 1, contribute to its earning power. Goodwill can be included in a balance sheet if it has been acquired for valuable consideration, almost always in a takeover, but not if it has been generated by the company itself. Since goodwill makes up a substantial part – often the majority – of the value of a company, particularly an innovative one, this rule has serious implications which will be discussed below.

Users of accounts look to the balance sheet for information which helps to estimate the risk of investing in it. Heavy reliance on short-term borrowing, as opposed to long-term capital, which results in what is known as a high gearing ratio, renders the company vulnerable to both rises in interest rates and demands for repayment of loans, with bank overdrafts being theoretically repayable on demand. The availability of cash to pay any such demands or other similar calls is also made clear by the balance sheet.

The profit and loss account

The minimum content of the profit and loss account, as specified by company law, is shown in Table 2.3. The law permits two essentially different forms of profit and loss account, both of which are shown in Table 2.3 in a way which reveals their common features and differences. In practice, the differences are minimised, because companies which choose the 'functional' form of profit and loss account (shown in the left-hand column of Table 2.3) – which divides costs into those of sales, distribution, and administration – have to show the two most important missing items from the 'value added' form (shown in the right-hand column of Table 2.3) – the total depreciation and amortisation costs, and the staff costs – separately in notes to the accounts. The only item of interest in the context of innovation is that for amortisation of intangible fixed assets, such as patents and other intellectual property, and of capitalised development costs.

Depreciation, which is charged on tangible assets, and amortisation, which is the preferred term for the equivalent charge on intangible assets, allow for the limited life of those assets. A machine, a patent, or a lease decline in value as time goes on, and this decline is effectively a cost of operating during the

Table 2.3 Profit and loss account formats

	Format 1	Format 2	
1	Turnover		1
2	Cost of sales	Change in stocks and finished	
3	Gross profit or loss	goods and in work in progress	2
4	Distribution costs	Own work capitalised	3
5	Administrative expenses		
6	Other operating income		4
		Raw materials and consumables	5(a)
		Other external charges	5(b)
		Staff costs:	6
		(a) Wages and salaries	
		(b) Social security costs	
		(c) Other pension costs	
		Depreciation and other amounts written off tangible and intangible fixed assets	7(a)
		Exceptional amounts written off current assets	7(b)
		Other operating charges	8
7	Income from shares in group undertakings		9
8	Income from participating interests		10
9	Income from other fixed asset investments		11
10	Other interest receivable and similar income		12
11	Amounts written off investments		13
12	Interest payable and similar charges		14
13	Tax on profit or loss on ordinary activities		15
14	Profit or loss on ordinary activities after taxation [Minority interests]		16
15	Extraordinary income		17
16	Extraordinary charges		18
17	Extraordinary profit or loss		19
18	Tax on extraordinary profit or loss [Minority interests]		20
19	Other taxes not shown under above items		21
20	Profit or loss for the financial year		22

Source: Companies Act 1985, as amended; presentation: DWB
Notes: Format 1 is the 'functional' version, while Format 2 is the 'value added' version. There are also Formats 3 and 4, which rearrange the contents of Formats 1 and 2 respectively into Charges and Income groups. As with the balance sheet formats, most companies seem to prefer the less logical version in favour of one or other of the versions given above, which provide more of a narrative.
   The references to minority interests are required in accounts for groups of companies.

period. The balance sheet value is thus reduced and the reduction is charged to the profit and loss account. The concept is superficially straightforward, and practice has become standardised, but there are some complex issues of both principle and practice which have a special significance for innovation, and which will be discussed below.

Since the costs of research cannot be capitalised, they must therefore appear as a charge for the year during which they are incurred, with no matching income item to set against them in most cases. (Some companies, particularly in the defence field, receive subsidies for research from public sources. Such subsidies appear as income.) Research is therefore done at the direct expense of profit in the year in which it is performed. As noted above, the same is true for development for the majority of companies.

In the absence of capitalised development expenditure, the Companies Act does not oblige companies to disclose a figure for research or development expenditure. It was argued strongly by some at the time of the Companies Act 1981 that the information was commercially confidential, and that such disclosure would do harm, particularly to small companies competing with larger ones. It was also feared that a requirement to reveal R and D expenditure would inevitably lead to subsequent demands for greater disclosure of information about the type and focus of the R and D. The legislators accepted some of these arguments and contented themselves with inserting a provision that, in their annual report, the directors should include:

> ... an indication of the activities (if any) of the company and its subsidiary undertakings in the field of research and development ...

Subsequently, however, concern about the shortcomings of British industry focused, amongst other things, on its alleged failure to spend or, as the advocates would say, 'invest' in R and D. It was argued that disclosure of R and D expenditure would shame the laggards into spending more. Faced with the implied threat that, if the accounting profession did not do something, the government would use its Order-making power to enforce disclosure, SSAP 13 was modified to impose such a requirement on public companies above ten times the size defined as a medium-sized company. Under the revised SSAP 13 (1989), such companies are obliged to state and explain their policy on R and D expenditure; and to disclose the total amount of expenditure on R and D charged to the profit and loss account – analysed between current expenditure and amortisation of deferred development expenditure – and the movements of the deferred expenditure, as well as the total of deferred development expenditure as provided for under law.

Users of accounts look to the profit and loss account for various measures of efficiency, such as employment costs to sales (shown as turnover in European and revenue in US accounts); for the 'interest cover', which is the ratio of profit before interest to interest payable; for the 'dividend cover', which is the ratio of profit to dividends; and possibly for an estimate of how much the company is spending on its future through R and D. In conjunction with the balance sheet, it provides the information to calculate such ratios as the return on net assets (RONA), return on capital employed (ROCE), or return on average capital employed (ROACE).

Z-scores

The various financial ratios derived from the balance sheet and the profit and loss account provide the experienced observer with a multi-dimensional financial view of an admittedly complex financial reality. Since the indications given by the different ratios may conflict, with some possibly suggesting robust health while others point to imminent decline, some method of combining them into a single figure has obvious attractions. Creditors in particular want advance warning of financial failure due to lack of cash, and the Z-score has been developed to meet this need.

The formulae used for calculating Z-scores are derived by statistical analysis of samples of companies which have failed and of those which have not, in order to find the formula which most precisely discriminates between the two groups, by producing negative Z-scores for the failed companies and positive values for the survivors. The formula can then be used on current companies. The particular formula used to calculate a Z-score has often been regarded as a commercial secret, but Taffler (1995) has given a useful review with some formulae. Taffler stresses that a negative Z-score is a necessary condition for subsequent failure, but not a sufficient one.

Examples of reporting of R and D expenditure

Although the minimum content of company accounts is prescribed by law and standards, there is a considerable degree of freedom about the presentation of the information. In particular, much of the detailed information can be included in notes rather than on the face of the accounts, which are often confined to the main features. Some illustrations of the different ways in which individual managements choose to present their R and D expenditure are given below.

The Companies Act requires that any figure given in the balance sheet or profit and loss account for the year being reported should be accompanied by the corresponding figure for the previous year. This provision ensures that the user of accounts can obtain some feeling for the trend of development within a company. Companies often include five-year or ten-year summaries of the principal figures from their accounts in a separate table, thus giving a better indication of the development of the company.

Rolls-Royce, the aerospace and power-generation equipment manufacturer, regards R and D expenditure as being one of the important items to be shown on the face of the profit and loss account, as shown in Table 2.4.

Table 2.4 Rolls-Royce plc: reporting of R and D expenditure

Extract from group profit and loss account:

	1994 £M	1993 £M
Turnover	3,163	3,518
Cost of sales	(2,646)	(2,995)
Gross profit	517	523
Commercial, marketing and product support costs	(117)	(105)
General and administrative costs	(91)	(89)
Research and development (net)	(218)	(253)
Operating profit	91	76

Source: Annual report 1994

Table 2.4 illustrates the use of the convention, common in published accounts, that negative figures are shown in brackets.

Rolls-Royce operates in an industry where major R and D projects are rare. Its R and D expenditure thus tends to follow the lumpy pattern of an individual project, with significant changes from year to year, rather than being smoothed by the different incidence of a range of projects at different stages in their lives. In the 1993–94 period, Rolls-Royce was in the later stages of the development of the Trent engine, and its R and D expenditure declined by £35 million (about 14 per cent) between the two years. It can be seen that the increase in its operating profit was only £15 million.

BOC, a company with a turnover very close to that of Rolls-Royce, is in the industrial and speciality gases business which is less demanding in R and D terms than aerospace. BOC presents its R and D expenditure as part of a note to the accounts (Table 2.5).

Table 2.5 The BOC Group plc: reporting of R and D expenditure

Note 2 to the Financial Statements:

	Continuing	Acquisi- tions	Before Except ional items	Except- -ional items	1994 Total £M
Cost of sales	(1,839.7)	(12.6)	(1,852.3)	(30.2)	(1,882.5)
Net operating expenses					
Distribution costs	(300.7)	(0.1)	(300.8)	—	(300.8)
Administration costs	(653.4)	(5.4)	(658.8)	(50.6)	(709.4)
R and D	(83.8)	—	(83.8)	(4.2)	(88.0)
Profits of related undertakings	38.1	—	38.1	—	38.1
Income from other fixed asset investments	0.7	—	0.7	—	0.7
	(999.1)	(5.5)	(1,004.6)	(54.8)	(1,059.4)

Source: Annual report 1994

SmithKline Beecham, the UK-US health care company, follows Rolls-Royce in including R and D expenditure in its profit and loss account, but also gives more detail in a note (Table 2.6).

Table 2.6 SmithKline Beecham plc: reporting of R and D expenditure

Extract from Profit and Loss Account:

	Comparable businesses £M	Acquis- itions £M	Business performance £M	One-off items £M	1994 £M
Sales					
Continuing operations	5,877	194	6,071	—	6,071
Discontinued operations	421	—	421	—	421
Cost of goods sold	(2,109)	(95)	(2,204)	(243)	(2,447)
Gross profit	4,189	99	4,288	(243)	4,045
Selling, general and					
administrative expenses	(2,261)	(85)	(2,346)	(320)	(2,666)
R and D expenditure	(620)	(1)	(621)	(17)	(638)
Trading profit					
Continuing operations	1,224	13	1,237	(580)	657
Discontinued operations	84	—	84	—	84

Note 2 to the Financial Statements: Trading Profit:

	Total 1994	One-off items 1994	Continuing operations 1994	Discontinued operations 1994	Total 1993
Cost of goods sold	2,447	243	2,044	160	2,065
Net operating expenses					
Distribution costs	181	—	173	8	178
Administration expenses	617	59	524	34	542
Other operating expenses	1,868	261	1,512	95	1,624
R and D expenditure	638	17	581	40	575
	3,304	337	2,790	177	2,919

The following amounts have been charged in arriving at trading profit:

	1994 £M	1993 £M
Amortisation of intangible fixed assets		
Goodwill	22	—
Licences, patents, etc	7	1

Source: Annual report 1994

SmithKline Beecham was formed by a transatlantic merger, and is quoted on the stock exchanges of both countries. It provides a note in its accounts reconciling the UK version to the figures which would have been produced under US GAAP (generally-accepted accounting principles). Under UK GAAP, its profit (net income) attributable to shareholders was £72 million for 1994: under US GAAP, it was £600 million, a striking demonstration of the effect of accounting conventions on reported profit. The difference in the balance sheets was equally marked, with the value of shareholders' funds being £570 million under UK GAAP, and £4,769 million under US GAAP. The differences arise largely from differences in the accounting treatments of goodwill and other intangible assets, a subject which is discussed at the end of this Chapter.

By contrast with SmithKline Beecham, we may look at Glaxo (Glaxo Wellcome from May 1995), the pharmaceutical manufacturer which is the UK's largest spender on R and D. Glaxo prefers the 'value added' format of its profit and loss account to the 'functional' format used by the other examples, and therefore cannot incorporate its R and D expenditure within the framework of the profit and loss statement. The value added format does, however, provide information about the composition of costs which reveals the cost structure of the business (Table 2.7). This information is helpful in financial modelling of companies, a topic taken up in Chapter 6 below.

Table 2.7 Glaxo Holdings plc: reporting of R and D expenditure

Extract from Note 3 to the Financial Statements:	1994 £M	1993 £M
Operating costs less other income:		
Raw materials, consumables and goods for resale	405	350
Staff costs	1,300	1,187
Depreciation	282	225
Other operating charges	1,864	1,649
Change in stocks of finished goods and work in progress	6	23
Share of losses (profits) of associated undertakings	3	(3)
	3,860	3,431
Less:		
Own work capitalised	12	13
Other operating income	11	13
	3,837	3,405
Operating costs include:		
Research and development expenditure	858	739

Source: Annual report 1994

Finally, British Aerospace (Table 2.8) provides an example of a company which defers development costs, which in this industry are included within the concept of launch costs. Launch costs comprise design and development, education, and jigs and tools. For projects with external funding, the practice in British Aerospace is to defer the launch costs and amortise them by reference to an assessment of sales.

Table 2.8 shows how the accounts of British Aerospace also disclose the substantial contribution to its total R and D expenditure which is made by others, together with the amount which it spends on training.

The British Aerospace accounts contain another interesting sidelight on the effect of accounting standards on the figures presented. The amount given in the 1993 accounts for finished goods and goods for sale was £575 million, but that given in the 1994 accounts for 1993 (as the previous year) was £1,119 million. The difference was explained as due to the requirements of the new accounting standard (FRS 5 1994), which required that recognition be deferred of some motor vehicle sales made by the Rover car group (sold by British Aerospace in March 1994) under stock financing arrangements.

Table 2.8 British Aerospace plc: treatment of deferred costs

Note 14; Stocks:

	Group 1993 £M	Group 1994 £M
Raw materials and consumables	155	95
Work in progress	3,306	3,516
Finished goods and goods for resale	1,119	285
Development properties	662	699
Progress payments	(1,350)	(2,035)
	3,892	2,560

Included within work in progress of the Group are unamortised launch costs related to programmes supported by launch aid of £501 million (1993: £486 million) and net costs less foreseeable losses on long-term contracts of £33 million for the Group and Company (1993: £242 million for the Group). The payments on account applicable to these contracts and included within progress payments are £33 million for the Group and Company (1993: £242 million for the Group).

Development properties for the Group include capitalised interest of £42 million (1993: £39 million) and rationalisation costs of £337 million for the Group (1993: £343 million) and £287 million for the Company (1993: £293 million).

Note 3; Operating costs (extract):

	1993	1994
Research, design and development		
Company funded	168	99
Other	442	344
Training	71	40

Source: Annual report 1994

OTHER FINANCIAL STATEMENTS

The balance sheet and profit and loss account do not reveal full information about the financial affairs of companies, and several other forms of financial reporting are in being. Two of these, the cash flow statement and the statement of recognised gains and losses, are required by the accounting standards FRS 1 (1991) and FRS 3 (1993), while the value added statement is a useful rearrangement of the information in the profit and loss account. These forms will now be briefly described.

The cash flow statement

Cash is the life-blood of a business and it is lack of cash, rather than lack of profit, which causes businesses to fail. A management which has ample amounts of cash at its disposal is much less constrained than one which has not. As various examples given so far illustrate, interpretation of the figures for profit and other measures of financial performance is not easy. There is, in consequence, increasing interest amongst investors in a company's ability to generate cash as being a better measure of its long-term performance than many other measures.

One reason for the interest in cash flow is that companies can, if so minded, manipulate published profit figures to show themselves in a favourable light. Detailed analyses of some cases are given by Smith (1992). Profit is, in any case, a subjective matter of opinion within certain limits. There can be legitimate differences of opinion about the allocation of some costs between current and future periods, about the value of stocks and work in progress, about the appropriate depreciation and amortisation rates to use, and about many other items in the accounts. Cash is, however, much more of an objective reality than profit.

A company's freedom to exercise its judgement is circumscribed by accounting standards, but these differ between countries with, as in the example of SmithKline Beecham, the same set of basic accounting figures giving rise to markedly different profit figures in different countries. Companies whose shares are traded on both the London and the New York Stock Exchanges therefore include in their published reports a reconciliation between the accounts drawn up according to the respective GAAP in the two countries.

An illustration of the variations within the European Union was provided by Simmonds and Azières (1989), who drew up a simplified profit and loss account and balance sheet from the recently-published accounts of a multinational group, together with a series of transactions designed to highlight areas where accounting treatment might differ. Accountants in each of seven countries were then asked to prepare accounts on the basis of the most likely accounting assumptions to be used in their country, and on alternative permissible assumptions which would produce net profit as high as, and as low as, possible.

The results of this comparison showed that the UK produced the highest most likely figure for net profit, at 192 (million ECU), while the average was 151 and the spread was from a minimum of 131 in Spain. The UK showed the least variation between the highest and lowest figures, with the maximum possible being 194, very close to the most likely value of 192, and the minimum 171. Germany, at the other extreme, showed a most likely figure of 133, with a maximum of 140 and a minimum of only 27.

Apart from deliberate manipulation, there are two main reasons why cash and profit may be different; timing differences and changes in the level of operations. Under the accruals principle, a profit is achieved when the relevant transactions have taken place, whether or not the relevant money has changed hands. Since the company making the profit will, under the normal terms of trade, owe less than it is owed, the profit will appear before the cash. Retailers, however, are in the fortunate position of receiving cash from their customers before they pay for their supplies, so in their case the cash appears before the profit.

In a company which is operating at a constant level, the profit will tend to show itself in cash, but operation at a constant level is unusual, with changes in that level being much more likely. The influence of inflation and devaluation on cash availability were illustrated in Chapter 1, but expansion of the company produces similar strains. It is for this reason that small companies often fail when they try to expand in response to favourable reception of their products, and why the level of company failures tends to rise at the beginning of an upturn in the economy. In such cases the company may be genuinely profitable but fail because it cannot pay its bills.

The cash flow statement is a way of illustrating these effects. It combines elements from the balance sheet, such as purchases of capital items, with items from the profit and loss account, and is based, not on the accruals principle, but on real cash flow from payments and receipts during the period to which it relates. It thus contains some information which is not available from the balance sheet and the profit and loss account.

The accounting standard for cash flow statements, FRS 1 (1991), replaced an earlier standard on source and application of funds which had been intended to reveal information about company liquidity but which had not proved to be as helpful as had been hoped. FRS 1 does not apply to companies which fall within the definition of small and medium-sized companies. The standard contents of the cash flow statement are shown in Table 2.9.

Table 2.9 The cash flow statement

Net cash inflow (outflow) from:
Operating activities
Returns on investments and servicing of finance
Tax paid
Net cash inflow (outflow) from investing activities
Net cash inflow (outflow) before financing
Net cash inflow (outflow) from financing
Increase (decrease) in cash and cash equivalents

Source: Company practice in applying FRS 1 (1991)

FRS 1 also requires that a note be provided to reconcile the operating profit and the cash flow from operating activities. One of the items in this reconciliation is the depreciation charge which is not a cash item but a notional charge. It adds to the cash reserves out of which investments, such as replacements for the depreciated assets, are made.

Another variation on the theme of the importance of cash can be seen in some biotechnology companies which cheerfully report losses from year to year, in accordance with plan, but remain in business because they have large cash reserves provided by their hopeful investors. The interest on these reserves may be a substantial contributor to the company's income. For reasons which will be discussed below, some such companies have been able to obtain a stock exchange listing before making a profit, thus making their accounts easily accessible. British Biotech plc is one of these companies, and its cash flow statement is shown in Table 2.10.

Table 2.10 British Biotech plc: cash flow statements

	1995 £000	1994 £000	1993 £000
Net cash outflow from operating activities	(26,276)	(20,377)	(14,754)
Returns on investments and servicing of finance			
Interest received	1,698	2,207	4,784
Interest paid	(479)	(492)	(492)
Interest element of finance lease payments	(76)	(63)	(98)
Net cash inflow from returns on investments and servicing of finance	1,143	1,652	4,194
Investing activities			
Purchase of tangible fixed assets	(7,131)	(3,612)	(3,979)
Sale of tangible fixed assets	16	7	8
Disposal of discontinued operation	—	879	—
Purchase of short-term investments	(44,900)	(18,500)	—
Sale of short-term investments	20,400	12,000	—
Net cash outflow from investing activities	(31,615)	(9,226)	(3,971)
Net cash outflow before financing	(56,748)	(27,951)	(14,531)
Financing			
Issue of shares and warrants	46,032	46	30,238
Expenses paid in connection with share issues	—	—	(2,617)
Repayment of loan	(270)	—	—
Capital element of finance lease payments	(77)	(218)	(233)
Net cash inflow (outflow) from financing	45,685	(172)	27,388
Increase (decrease) in cash and equivalents	(11,063)	(28,123)	12,857
Cash utilised by operations			
Increase (decrease) in cash and equivalents	(11,063)	(28,123)	12,857
Net purchase of short-term investments	24,500	6,500	—
(Inflow) outflow from financing	(45,685)	172	(27,388)
Cash utilised by operations	(32,248)	(21,451)	(14,531)

Source: Annual reports 1994, 1995

The statement of recognised gains and losses

The purpose of this statement is to reveal gains or losses which do not pass through the profit and loss account, thus upsetting the established relationship between the two. These gains or losses are primarily revaluations and exchange rate adjustments, and are not of direct interest in the context of innovation. The reasoning behind their existence, however, is very relevant to innovation, for it goes to the heart of the issue of maintaining the wealth of the company. These issues are taken up again at the end of this chapter.

The value added statement

The value added statement was a popular voluntary, unaudited, addition to company reporting in the late 1970s, its virtues having been strongly advocated in an influential discussion paper 'The Corporate Report' (ASSC 1975). This document says:

> The simplest and most immediate way of putting profit into proper perspective *vis-à-vis* the whole enterprise as a collective effort by capital, management and employees is by presentation of a statement of value added (that is, sales income less materials and services purchased). Value added is the wealth the reporting entity has been able to create by its own and its employees' efforts.

The value added statement became less popular after the 1981 Companies Act, whose requirements ensured that the information included in it would be available, albeit in a less convenient form, and it has almost disappeared.

The value added statement rearranges items from the profit and loss account into a form which, as ASSC (1975) pointed out, puts the various inputs and outputs into perspective. In the short run, a company cannot alter the amount which it receives from sales and the amount which it has to pay out to suppliers. The difference between these two sums – the value added – is the amount which the management has at its disposal to pay the employees, service its capital by dividend and interest, provide for depreciation, pay its taxes and, if all goes well, retain for future investment.

One of the reasons for the initial popularity of the value added statement was as a vehicle for communicating with employees who were expected to find it simpler to understand than a profit and loss account. This simplicity is, however, not obtained at the expense of usefulness, and analysis of the distribution of value added tends to be more revealing about a company's strategic direction and prospects than is sometimes appreciated. In particular, too high a share of value added devoted to employee costs is a good leading indicator of company failure, a notable example being that of the British automobile industry whose virtual collapse in 1974 was predicted in 1972 on the basis of analysis of value added figures for 1970. Another example is that of Dunlop which as early as 1979 published a statement which revealed that 88 per cent of its value added was devoted to employment costs, 8 per cent to interest on loans, and 5 per cent to taxes. Despite thus devoting 101 per cent of its value added to merely keeping going, it was four years later, in 1983, before the company finally collapsed.

With the renewed interest in the relations of a company with all its 'stakeholders', stimulated particularly by the 'Tomorrow's Company' Inquiry

(RSA 1995), the value added statement may again become popular. It is in any case possible to reassemble a value added statement from the accounts of a company but, as ICI continues to include one as part of its annual report, we can use that as a foundation for an analysis in stakeholder terms, as shown in Table 2.11.

Table 2.11 ICI plc: stakeholder shares of value added, 1991

	£M	% sales	%VA	Stakeholder
Sources of Income				
Sales turnover	12,488			
Royalties, etc	119			
Total	12,607	100		Customers
Less				
Materials, etc used	(8,285)	65.7		Suppliers
Value added	4,322	34.3		
From related companies	30	0.2		Associates
TOTAL VALUE ADDED	4,352	34.5	100	
Disposal of Value Added				
Pay, pensions, etc	(2,726)			
Profit sharing bonus	(31)			
Total employment	(2,757)	21.9	63.4	Employees
Taxes	(279)			
Less grants	17			
Net taxes	(262)	2.1	6.0	Government
Providers of Capital:				
Net interest	(220)	1.7	5.0	Lenders
Dividends	(391)	3.1	9.0	Shareholders
Minority shareholders	(22)	0.2	0.5	Associates
Total to providers	(633)	5.0	14.5	
Re-investment:				
Depreciation	(549)	4.3	12.6	
Retained profit	(151)	1.2	3.5	
Reinvestment	(700)	5.5	16.1	Company
ADDED VALUE DISPOSED	(4,352)	34.5	100	

Sources: Data: Annual report 1991; calculations and annotations: DWB

Innovation expenditure in the value added statement

In the form provided by ICI, the value added statement does not give any information relevant to innovative activity, but it could be readily adapted to do so, and is in many ways ideal for the purpose, being unregulated by law or accounting standards, and therefore available for experimentation by companies anxious to improve their communication with their investors and other interested parties, and for internal strategic management purposes.

This argument, and indeed the value added presentation generally, seems to appeal much more to engineers and scientists who think in terms of systems and their working than it does to accountants and economists whose concern is more with outcomes. For a few years, Amersham International, under a technically-trained Chief Executive, published a value added statement which showed R and D expenditure as an integral item. The company ceased to publish a value added statement following the succession of an accountant to the post of Chief Executive. The figures for the last four years in which it was published are shown in Table 2.12.

Table 2.12 Amersham International plc: R and D expenditure in value added statements

Year Ended 31 March:	1986 £000	1987 £000	1988 £000	1989 £000
Turnover	119,760	148,491	165,010	179,979
Bought-in materials, services and depreciation	51,554	64,421	71,961	80,472
Value added	68,206	84,070	93,049	99,507
Applied as follows:				
Employees:				
Wages, salaries and pension scheme costs	42,350	52,231	56,831	63,311
less: research and development labour included below	6,479	7,505	8,620	9,321
[net]	35,871	44,726	48,211	53,990
Governments:				
Corporate taxation in the UK and overseas	4,699	7,097	8,059	7,216
Providers of capital:				
Interest on loans	2,148	2,360	2,837	3,802
To partners in companies not wholly owned	587	1,633	736	1,302
Dividends to shareholders	3,507	4,130	5,058	5,619
Total	6,242	8,123	8,631	10,723
Reinvested in the business:				
Retained profit	8,761	9,220	11,421	7,301
Research and development	12,633	14,904	16,727	20,277
Total	21,394	24,124	28,148	27,578
Total application of value added	68,206	84,070	93,049	99,507

Source: Annual reports 1987, 1988, 1989

The Amersham value added statement usefully illustrates two points. First, Amersham deducted depreciation before arriving at a figure for value added. This treatment, which some other companies also adopted, is justifiable in economic terms, but has the disadvantage of introducing a subjective element into the valuation. The problem is, however, not a serious one for the user of accounts because the depreciation provision will be found in the accounts and can be added back.

The second point illustrated by the Amersham figures is that care has to be taken over double counting, especially of employment costs, which have to be reported as an item, but which constitute a substantial proportion (about half in Amersham's case) of the total R and D expenditure. It is paradoxical that while a value added statement lends itself very well to presenting R and D expenditure (and innovation expenditure in general) in context, the 'value added' format of the profit and loss account permitted by the Companies Act lends itself less readily to such disclosure than does the functional form. It would be possible to show the analysis of employment costs into R and D (or total innovation) and other categories in a 'value added' profit and loss account since there is nothing to stop companies giving more detail than the law and accounting standards require, but it is hardly surprising that they do not do so.

Statement of accounting policies

A company's accounts have to include a statement of the accounting policies which have guided their preparation. These policies will include the accounting treatment of R and D. This statement is usually found with the notes to the accounts which usually contain the detailed information required by the Companies Act, with only abbreviated versions of the main statements being shown as the main items. As Smith (1992) recommends, a user of accounts is well advised to start with the notes and work backwards through the usual order of presentation.

Companies are permitted to circulate only an abbreviated version of their accounts, lacking the notes and other detail, to shareholders who are content to receive such an inadequate description of the repository of their investment.

Directors' report

The directors must report each year on their conduct of the business, and are obliged to include within this statement 'an indication of any activities in the field of research and development'. It will be remembered from above that this requirement predates that of SSAP 13 to disclose a figure for expenditure, on which the directors' report therefore acts as a commentary. The importance given to this item by individual companies varies, as would be expected from the variation in importance of R and D from sector to sector of the economy.

THE PURPOSE OF ACCOUNTS AND ISSUES OF PRINCIPLE

The purpose of accounts

One of the guidelines followed by the ASB is of ensuring that '... the information resulting from the application of accounting standards faithfully represents the underlying commercial activity' (ASB 1991a). The ASB defines the objective of financial statements as 'to provide information about the financial position, performance, and financial adaptability of an enterprise that is useful to a wide range of users in making economic decisions' (ASB 1991b). The ASB envisages that users will wish to evaluate the capacity of the reporting company to generate cash in order to pay its employees and suppliers, service its finance, and make the investments which will enable it to stay in business, and can thus be seen to take a similar view to that of modern investment theory with its heavy emphasis on cash (Chapter 1).

Other than the figure for cash in hand and at the bank – which is true only for the day to which it relates – there is good deal of scope for judgement about the values to be put on items in financial statements. It is the purpose of accounting standards to limit the scope of this judgement in the interests of producing a true and fair view of the results of an individual company which will, at the same time, be reasonably suitable for comparison with the results of other companies.

An underlying, and so far unresolved, problem is, however, that users of accounts may have different interests. Accounts, after all, reflect the past, and their value as guides to the future must be limited. In practice, accounts are still heavily influenced by their history as the means by which directors reported on their stewardship of the shareholders' funds, and on how much profit had been made for distribution to those shareholders.

The figure for profit is also of interest to the tax authorities who, in the UK, the US, and the Netherlands, require that accounts be drawn up for their purposes according to their rules, which may be different from those adopted by the directors for the presentation to shareholders. In Japan, Germany, France, and other member states of the European Union, companies normally use the same accounts for both tax and shareholder purposes.

Accounting profit and economic profit

The relationship between accounting profit and economic profit is a complex one (Edwards, Kay and Meyer 1987), the details of which are beyond the scope of this book. The main issues are, however, of considerable relevance to our topic of innovation because the underlying concept of economic profit is the sustainability of the business, to which innovation nowadays makes the principal practical contribution. In current accounting practice this contribution is imperfectly acknowledged. Some guidance on possible lines of development is, however, provided by the issues of principle which emerge when considering other aspects of maintaining the sustainability of the business.

It will be remembered from Chapter 1 that the definition of the economic profit of a company for a period is the maximum amount which can be taken out of the company at the end of the period while leaving it as well off as it was at the beginning of that period. This definition is straightforward but the definition of 'as well off as' is not. At its most basic it could be taken to mean 'having the same capital', that is, the value of the capital put in by the shareholders remains unimpaired. Since some of this capital will have been used to buy fixed assets, such as machinery and buildings, maintaining the value of this capital involves making adjustments, through the depreciation provision, for the diminution of the value of those assets with time. This principle is well established and, since 1981, has been enshrined in company law. This law leaves the choice of depreciation rate to the directors, but the tax authorities specify the rates which are acceptable for their purposes.

Accounting for inflation

Unfortunately, money does not always retain its value over time. Throughout recorded history, inflation has generally reduced the value of the currency unit, so that a pound, dollar, mark, franc, or other unit is usually worth less at the end of a period than it was at the beginning. Maintaining the capital of a company in currency units thus does not maintain its wealth in terms of purchasing power, and a second definition of 'as well off as' is therefore that of maintaining the purchasing power of its capital.

Inflation has been far from uniform over the years, and has been negative in some periods, most recently in the UK in the 1930s. For much of business history it has not been sufficiently great to call for much adjustment of accounts prepared on the basis of historic costs. From time to time, however, and particularly during the period 1970–1985 in the UK, it has been at a high enough level to ensure that profits calculated on a historical cost basis were unrealistically high and did not reflect the amount which could safely be removed from the company while still leaving it as well off in any sense as it had been.

Various proposals were therefore put forward for adjusting accounts to allow for the fall in the value of money. Two main bases were suggested: the constant purchasing power (CPP) method, under which all values were adjusted on the basis of a general price inflation index; and the current cost accounting (CCA) basis, under which values were adjusted to current costs for equivalent assets. The philosophical bases of these two forms of adjustment are quite different: one reflects the change in the value of money, while the other reflects changes in values as they affect the individual company.

There is a good argument for saying that both adjustments should be made but the ASC, when it issued a standard on inflation accounting (SSAP 16 1980), opted for CCA alone. SSAP 16 was issued for a trial period of three years for accounts for periods starting on or after 1 January 1980. Under its provisions, historical cost information was also to be provided, and by the end of 1983 it was clear that concerns about the reliability of the CCA figures which were produced under it, the continued use of historical cost figures for tax purposes, and a drop in the rate of inflation had effectively destroyed its acceptability to both preparers and users of accounts. From the investors' point of view, a further reason for lack of interest in CCA accounts was that the relative performance of different companies was not much affected. SSAP 16 was therefore officially suspended in mid-1985 and finally withdrawn in mid-1988.

Mumford (1979) has shown how the pattern of interest in inflation accounting over the previous inflationary period 1948–54 closely parallelled that of 1973–78. He distinguishes a cycle of eight stages: increase in domestic inflation and a fall in stock market prices; reaction by the accounting profession; intervention by government; 'radical' studies of accounting techniques; controversy in the profession; a compromise recommendation; a reduction in domestic inflation and a recovery of stock market prices; interest in reform dwindles.

The approach taken by SSAP 16 was, as mentioned above, to revalue historical figures or make other adjustments in line with current costs rather than with a general index of inflation on CPP lines. Much of the doubt about the reliability of figures produced under SSAP 16 arose from the subjectivity inherent in this choice. The adjustments required by SSAP 16 were as follows:

For the profit and loss account, to allow for the impact of price changes on the funds needed to maintain the net operating assets:

A depreciation adjustment to allow for the difference in the reduction of the value to the business and the historical cost depreciation. Value to the business was defined as the lesser of the net current replacement cost or the recoverable amount. Recoverable amount was in turn defined as the greater of net realisable value and the amount recoverable from further use. (See Chapter 3 for explanation of these terms.)

A cost of sales adjustment to reflect the difference between the the value to the business and the historical cost of stock used in the period.

An adjustment to reflect any increase needed in the monetary working capital, which was defined as the aggregate of trade debtors and similar items, stocks not subject to a cost of sales adjustment, and trade creditors and similar items related to the day-to-day working of the business.

Where the company had net borrowings, a gearing adjustment to reflect their diminution in current cost terms. This adjustment added to the profit if prices rose.

For the balance sheet, values were to be expressed at their value to the business or directors' valuation, depending on the asset. Current assets and all liabilities were to be valued on a historical cost basis.

As the example of the Amersham accounts (Table 2.13) shows, the differences between historical cost and current cost accounts were significant, especially for individual items. The adjustments do, however, cancel out to some extent, because, for example, loans become less valuable, as many a mortgaged home owner has discovered, offsetting the increase in provisions needed to maintain the value of assets.

The inflation accounting story also illustrates the subjectivity of accounts and the differences in possible methods of valuation of assets, a subject which we examine in more detail in the next chapter.

Table 2.13 Amersham plc: historic and current cost comparison (1984)

	Historic cost (£000)	Current cost (£000)
Profit and Loss Account:		
Turnover	87,583	87,583
Historic cost operating profit	14,604	14,604
Current cost operating adjustments		(2,038)
Current cost operating profit		12,566
Gearing adjustment		361
Interest	(879)	(879)
Profit before taxation	13,725	12,048
Taxation	(4,661)	(4,661)
	9,064	7,387
Profit attributable to minority shareholders	(873)	(858)
	8,191	6,529
Dividends paid and proposed	(2,500)	(2,500)
Retained profit	5,691	4,029
Balance Sheet:		
Fixed assets		
Tangible assets	49,025	66,586
Current assets		
Stocks	12,950	13,158
Debtors	19,764	
Short-term deposits	1,894	
Cash at bank and in hand	1,614	
Monetary working capital (net)		11,130
	36,222	24,288
Creditors – amounts falling due within one year		
Loans	9,964	
Other creditors	14,645	12,467
	24,609	
Net current assets	11,613	11,821
Total assets less current liabilities	60,638	78,407
Creditors – amounts falling due after more than one year	(4,980)	(4,980)
Provisions for liabilities and charges	(1,593)	(1,593)
Accruals and deferred income		
Investment grants	(3,730)	(5,125)
	50,335	66,709
Capital and Reserves		
Called-up share capital	12,501	12,501
Share premium account	399	399
Other reserves	1,744	
Current cost reserves		25,421
Profit and loss account	34,530	27,175
	49,174	65,496
Minority interests	1,161	1,213
	50,335	66,709

Source: Annual report 1984 (statements combined by DWB)

One legacy of inflation accounting is the fairly widespread practice of revaluing some fixed assets, particularly freehold and leasehold property. Revaluation of fixed assets is permitted under the alternative accounting rules (for inflation accounting) in the Companies Act. Each revaluation gives rise to an entry in the statement of recognised gains and losses, to an item under 'revaluation reserve' in the balance sheet to offset the increased value of the asset, and to an increase in the depreciation provision in the profit and loss account (SSAP 12 1987).

Maintaining the earning power of a company

There is a third definition of 'as well off as' which is particularly relevant to innovation, and this is 'having the same earning power'. In the early days of company law and the development of accounting, the earning power of a company was largely determined by its physical and financial assets, and maintaining their value essentially maintained the earning power of the company. As we have seen, this is no longer true, and the earning power of a modern company increasingly depends on intangible assets such as patents, and intellectual assets, such as knowledge, skills, and established reputation.

The value of a company's intellectual capital, as the total of its assets which are neither physical nor financial is increasingly being called, shows up in the difference between the value of the company on the stock exchange, determined by multiplying the share price by the number of shares in issue, and its value as shown in the balance sheet as shareholders' funds. The 'market to book' ratio of these values is being used by some companies as a measure of their performance in building up their intellectual capital.

The market to book ratio is an easily-calculated measure for tracking the trend of development, but it is likely to be inaccurate on two counts. First, the book value may not, indeed almost certainly will not, be an accurate reflection of the value of the company's tangible assets; second, the share price is that which brings together willing buyers and sellers of relatively few shares. More accurate values are provided when a company is taken over, with all or most of its shares being bought by another. The share price is usually higher in these circumstances than it was when only marginal amounts of shares were being traded. If the takeover is regarded as an acquisition, rather than a merger, the assets of the company taken over are revalued for inclusion in the accounts of the acquirer, producing more realistic values than those shown in the balance sheet of the acquiree.

The difference between the amount paid for a company and the value of the assets (including intangible assets) acquired is known as goodwill. A study by Higson (1990) showed that the goodwill element in takeovers had increased from around 1 per cent of the bidder's net worth in 1976 to about 50 per cent in 1987.

The goodwill or intellectual asset element of a company's value largely arises from its expenditure on training, organisation, management, R and D, advertising and other brand-building expenditure which, for reasons of accounting prudence which, it will be remembered, is a requirement of company law, are regarded as current expenditure to be written off in the year in which it is incurred. Nonetheless, this expenditure is different in kind and object from that on such items as raw materials and consumables which are genuinely current expenditure, since its main purpose is to provide the earning power of the future. This category of expenditure is at the discretion of management and is therefore sometimes known as 'discretionary expenditure'. An alternative description is 'revenue investment', a term which captures the accounting rather than the management aspect of its uncertain status.

Accounting for goodwill

Goodwill – defined in the relevant UK accounting standard, SSAP 22 (1989), as the difference between the value of a business as a whole and the aggregate value of the fair values of its separable net assets – poses difficult and as yet unresolved problems of accounting. SSAP 22 permits purchased goodwill to be dealt with in two ways. The preferred method is to write it off against reserves immediately on acquisition, thus in effect buying it out of retained profits. The alternative is to capitalise and depreciate it over its useful economic life, thus buying it out of future profits.

The reason for the preference for immediate write-off of purchased goodwill is that this treatment preserves equality of treatment with self-generated goodwill, which cannot legally be capitalised. Even if this were possible, however, there would be great difficulties in valuing self-generated goodwill in a manner which would command general acceptance. The argument for capitalisation and write-off is that it reflects the fact that the company has acquired a genuine asset – something which will generate future revenue – at an ascertainable price which the acquirer regards as fair. A further argument in favour of capitalisation and amortisation is that this is the preferred treatment in other countries, and that uniformity of treatment is desirable in an increasingly global marketplace for capital.

UK managements favour immediate write-off against reserves because it avoids affecting the future profit, to which the stock market pays considerable attention, and in some cases because they have a direct financial interest arising from a bonus scheme geared to profit. With the rise in the importance of the goodwill element in takeovers, arising from the increasing importance of intellectual and intangible assets as described in Chapter 1 and documented by Higson (1990), the effect on the balance sheet of immediate write-off, however, became so great in some cases as to become a potential embarrassment. This was especially a concern when the company was in danger of exceeding:

- a limit on the ratio of loans to shareholders' funds agreed with its bankers or other providers of loan capital;
- a similar limit on the freedom of the directors to borrow, prescribed in its Articles of Association;

or of having such a small amount of net assets that a comparatively small acquisition or disposal would be above the limit of 25 per cent of net assets above which the listing agreement of the London Stock Exchange requires the consent of the shareholders in general meeting to be obtained to the transaction.

Managements therefore began to look for ways in which what they considered to be a more realistic view of the business could be presented. One way was to find elements of goodwill which could be regarded as separable, intangible, and arguably non-depreciating assets whose value could be determined and put on the balance sheet. Brands in particular were identified from 1988 as satisfying this criterion, and methods of valuing brands and other intangible assets are discussed in Chapter 3 below.

Another way to correct the perception was to show the amount written off. The engineering company TI was one of the first companies to adopt this tactic, in its 1988 accounts, following a period in which it had undergone considerable restructuring, involving selling various businesses and acquiring others.

In its annual report, TI explained both the situation and its response to it as follows:

> Under UK law, goodwill arising on consolidation cannot be carried in the consolidated balance sheet without a requirement to depreciate it. For TI, depreciation of goodwill would be unnecessary, as substantial revenue sums were spent every year to maintain and enhance the competitiveness of the business, thus increasing the value of the intangible assets over time.
>
> Individually identified intangible assets can be carried in the consolidated balance sheet without having to be depreciated, provided that they are separable, that is, that they could be sold separately from the tangible assets.
>
> For an engineering group such as TI, with a high service content, the intangible assets are not separable since they include patents, tradenames, trained people, CAD/CAM and other systems, software and engineering drawings, all of which are integrated with the tangible assets to form the basis of the profit and cash-generating capacity of the company.
>
> Consequently, the directors believed that it would not be in the shareholders' interests to incur restrictions on the ability to use borrowings, within prudent limits, where to do so would enhance earnings per share. In so doing, however, the directors wished to avoid raising possible concern that the level of borrowings was so high in relation to the net tangible assets appearing in the balance sheet (a relationship known as the gearing ratio) that the company might have to seek further funds from shareholders.

In the consolidated balance sheet, the directors therefore showed shareholders' funds prior to the write-off of goodwill (Table 2.14), and said that this figure would be used as the basis for assessing TI's prudent debt capacity, in conjunction with interest cover.

Table 2.14 TI Group plc: treatment of goodwill

Extract from Balance Sheet:				1988 £M	1987 £M
TI shareholders' funds – gross				501.3	384.6
Goodwill written off				(268.8)	(180.3)
TI shareholders' funds – net				232.5	204.3
Interests of minority shareholders				3.4	6.0
Total shareholders' funds				235.9	210.3
Extract from Group financial history:					
	1984	1985	1986	1987	1988
Interest cover (times)	2.0	2.5	3.9	18.8	12.1
Gearing:					
gross	42.3%	33.0%	45.8%	Nil	8.1%
net	42.3%	33.0%	45.8%	Nil	17.8%

Source: Annual report 1988

The role of the balance sheet

The discussions on inflation accounting illustrate a basic problem with company accounts which is that they attempt to serve several purposes which have become less compatible as the importance of intellectual capital has increased, and asset values have become less stable under the influence of inflation and technological obsolescence.

It is generally acknowledged, not least by the ASB, that the result in the present state of the art of financial reporting is to rob the balance sheet of the clear meaning it may once have had when a company's assets were essentially what appeared on it. There are two schools of thought as to how the problem should be solved. One school wishes to make changes in the balance sheet to bring it more into line with the real assets so that it becomes a genuine measure of wealth. This school therefore favours revaluations of assets to current values and the inclusion of purchased intangible assets on the balance sheet.

The other group regards the balance sheet as no more than a record of residuals – amounts which have been expended in cash terms but which have not as yet been fully allocated as costs under the matching principle. This school regards the profit and loss account as being the more important of the two principal statements, whereas the former school allocates that status to the balance sheet.

The ASB appears to favour a gradual movement towards the primacy of the balance sheet and further possible moves in this direction are discussed in Chapter 7 below.

Reporting on revenue investment

As has been described above, the R and D component of revenue investment attracted sufficient attention to persuade the ASC to introduce a requirement of its disclosure in the accounts of larger companies. The ASC's successor, the ASB, attempted to extend this disclosure to the other elements of revenue investment which in FRED 1 (1991) it defined as R and D, training, advertising, and major maintenance and refurbishment expenditure.

The general reaction to this proposal was such that the ASB decided not to go ahead with it as a component of its definitive standard FRS 3 (1993). Instead, being still convinced that information on revenue investment would be of assistance to users of financial statements, the ASB proposed instead that such information be included within a new, voluntary, statement to be called the Operating and Financial Review (OFR) (ASB 1993). The concept of the OFR is as a framework within which the directors can discuss and interpret the progress of the business in order to put the financial statements into perspective. In the OFR, revenue investment is included with capital expenditure in a section on 'investment for the future'. Included in the examples of current expenditure for the benefit of the future are marketing and advertising campaigns, training programmes, and R and D.

The OFR gives scope for experiment with forms of presentation which could not possibly be provided within an accounting standard, and is at present the most likely vehicle for disclosure of reliable information about a UK company's innovative activities. The discussion of these possibilities is deferred until Chapter 7 below, but it is worth noting here that UK expenditure on advertising is comparable to that on R and D. For 1993, the respective figures were £9,160 million for advertising (NTC 1994) and £13,753 million for R and D (CSO 1995).

These innovative activities are, of course, a focus for internal management attention, not least in a financial sense, and it is to the topics of valuation and management accounting that we now turn.

Questions for discussion

1 The ASB believes that no single number can encapsulate the financial performance of a company, but Z-scores are believed to have some predictive value. Is there a conflict between these beliefs?

2 ICI spent £596 million on R and D in 1991, and another £96 million on technical service. What percentage of value added was thus used?

3 Would the development of reporting on innovation be better concentrated on the profit and loss account or the balance sheet?

4 Why is there so much interest in R and D spending while advertising expenditure attracts little interest, as do other revenue investments such as exploration expenditure by mineral companies or the costs of preparing bids for major capital goods contracts?

Chapter 3

Valuation

THE NEED FOR VALUATION

All business activity is ultimately judged against some standard, explicit or implicit, of value for money. The key question which any manager will ask when faced with a spending decision is: will what is likely to be gained by spending this amount of money be worth what has been spent? Sometimes the criterion for measuring the gain will be more strategic than financial where, for example, the expenditure is necessary to reduce or eliminate some source of pollution which, if continued, would result in the business being closed down. Assuming that the business wishes to continue, then the money must be spent. Even so, there is an implicit strategic judgement that the business will be worth continuing. At the other end of the strategic scale there may be a choice between two opportunities for new product development, only one of which can be afforded. In this case, although there may be strategic considerations which favour one project over the other, the decision will be strongly influenced by which opportunity is likely to give the higher return.

In both cases, and in those of intermediate difficulty, a necessary input to the decision is a valuation of the return.

Valuation is also important in issues such as licensing in or out of patent rights, know-how, or other IPR. Even if the transaction is between companies which are part of the same group which itself takes a fairly relaxed view of the valuation of assets for internal transfer purposes, the tax authorities may well become interested if the transfer is across borders in case there should be a possibility that the transfer is effectively being used to move taxable profits from one jurisdiction to another. An independent valuation may be required in such cases.

AVAILABLE TECHNIQUES FOR VALUATION

Techniques for valuing the intangible assets which are important in innovation decisions are based on well-established techniques for valuing real property such as land and buildings (Smith and Parr 1994). Although the principles are the same, there tends to be more scope for variation in the values of intangible assets than in the cases for which the techniques were originally developed, particularly as the range of possible outcomes is more varied.

It is important to recognise that value is not an absolute attribute of an asset. Value depends on the circumstances, as the substantial numbers of home owners who bought at the height of the 1980s UK property boom are only too well aware. Properties bought at a time when prices were rising rapidly and were expected to continue to do so, tax relief on mortgage interest was relatively generous, and buyers felt secure in their employment, dropped substantially in value as these factors changed under the influence of recession, tax changes, and increased insecurity of employment. The time, the circumstances of the parties to the transaction, and the economic environment are all major factors in valuation, and all are specific to the individual transaction.

An objective value is reached only when a willing and informed seller and a willing and informed buyer agree on a price which both consider fair and reasonable. Even this value is valid only at the specific time at which it was agreed. In the English system of property sales there is a delay between acceptance of an offer and its becoming legally binding, with the result that gazumping has sometimes occurred in times of rapidly-rising prices and its reverse during slumps. Such behaviour is unknown in the Scottish legal system in which an offer is binding on acceptance. Knowledge of the legal framework is a part of the information which parties to the contract need to have.

The value, as determined by such a sale of a defined property, is available only after the event. For those contemplating such transactions, or seeking to value assets for taxation or other reasons, valuation must be by considering a virtual transaction – one which would have been done had the opportunity offered. There are three basic approaches to making valuations in these circumstances based respectively on cost, market value, and likely return.

Over the last twenty years or so, techniques have been developed for valuing rather more nebulous items, particularly brands and financial derivatives such as options to buy shares at a specified price at a specified future date. Both have some relevance to innovation since brand values are largely the result of investment in intellectual assets, while a successful R and D project can be considered to provide an option to carry on to the exploitation stage of innovation, rather than as an investment.

The most popular technique developed for valuing brands produces a result which is closely similar to that for valuing shares in that it expresses the value as a multiple of earnings.

The technique for valuing options is based on recognition of two key concepts. One is asymmetry – the potential loss is limited to the cost of the option (or of the R and D project), but the potential gain may be anywhere within a more-or-less wide range. The second concept, which is much more reliably applied to share prices than to R and D projects, is that the future price or value range, and the probability of the gain lying at any point within it, are statistically determined. It is the existence of statistical behaviour which enables risk to be managed: if behaviour is completely unpredictable then we are dealing with uncertainty which is much more difficult to manage.

In this chapter we first look at the standard methods of valuation, then at the valuation of brands, and finally at the options approach and its applications.

Cost basis of valuation

The cost basis of valuation is that normally used for company balance sheets when it is known as the historic cost method. Assets are entered on the balance sheet at the price that was paid for them. Accountants like this method because it is auditable. The price paid, and hence the figure in the balance sheet, can be checked against invoices or similar documents or records.

As a record of what the directors have done with the shareholders' money, the historic cost method is unarguably correct. The figure is arrived at objectively with no element of judgement involved.

The objectivity does, however, not last long. The asset, unless it is land, will be expected to decline in value as it is used for the purposes for which it was bought, thus wearing out over a period known as its useful life. If the balance sheet is to give a true and fair view, the value attributed to the asset must be reduced, or depreciated, to zero over this useful life. Normal accounting practice is to make this reduction by equal instalments over intervals of time, of which the annual accounting period is the most significant. This approach is known as the 'straight line' method of depreciation. The value shown on the balance sheet is known as the 'book value' or 'net book value' of the asset. An alternative method, known as the 'reducing balance' basis of depreciation, is to depreciate the book value by a constant proportion each year.

The depreciation allowance must cover not only physical deterioration, but also technological obsolescence or other changes which make an asset worth less, or occasionally more, than its depreciated historical cost. Keeping up to date in this way is just as important to maintaining a company as a going concern as is making good ordinary wear and tear.

The depreciation policies followed by a company are reported in its annual accounts in the section on accounting policies. One of the stratagems adopted by companies in trouble is to change depreciation policy in order to reduce depreciation charges, and thus increase profits, so that the user of accounts is well advised to check this point.

The balance sheet deals only with the capital side of the business. To see what has been gained from the consumption of capital arising from depreciation, we have to look at the profit and loss account. The usual straight line method effectively allocates the cost of the asset uniformly over its estimated period of use, and the amount of depreciation in a period is effectively a cost of production over that period. This amount is therefore charged to the profit and loss account. While real in the sense that it represents a diminution in the company's total assets, it is not an expense in the sense that money goes out of the company during the period, as it does for the purchase of raw materials or the payment of wages. It rightly comes off the profit, but remains as cash in the business, effectively replacing the cash value lost by the wearing out of the original physical asset during the period.

It was mentioned in Chapter 2 that the depreciation charge for financial accounting purposes may be different from the figure for tax purposes. Although, sooner or later, the company's depreciation provisions and those allowed for tax purposes must be reconciled, that point can be long postponed in practice. Meanwhile, the tax treatment has real effects on the company, particularly on its cash flow. Capital assets must be bought and paid for out of available funds, and only the proportion of that cost allowable for depreciation for tax purposes will be treated as an expense against income in the year of purchase. The rest of the purchase cost must be found from reserves or after-tax income. It is sometimes argued that this treatment inhibits desirable investment, especially in periods of recession, and tax authorities occasionally respond by temporarily increasing allowable depreciation, sometimes to as high as 100 per cent in the first year.

In the UK, capital equipment bought for research purposes is 100 per cent allowable for tax. Otherwise, the Inland Revenue applies standard rates of 25 per cent on a reducing balance basis to plant and machinery, motor cars, and patent rights and know-how; and 4 per cent on a straight line basis to buildings.

Time brings depreciation, but it also tends to bring inflation. Historical cost depreciation over the useful life of an asset will thus enable the shareholders to recoup their original money, but it will be worth less in purchasing power than when they invested it. Only in the unusual circumstances of falling prices, such as those which currently apply to most electronic equipment, and particularly to computers, will the accumulated depreciation provide enough money to buy an equivalent replacement asset. As this is what is required to keep the business in the same real operational state as it was when it began, other valuation techniques, such as CPP and CCA which allow for inflation effects, are therefore advocated from time to time for use in balance sheets, and are widely used for other valuation purposes.

Before turning to some of these alternative techniques, we can summarise that the cost-based valuation techniques must also include consideration of depreciation and inflation.

Fair market basis of valuation

The 'fair market' or 'arm's length' value is the ideal, arrived at by a willing and informed seller and a willing and informed buyer. Rather than a balance sheet value, it can be thought of as a value for insurance purposes. It is a current valuation, adjusted normally on an annual basis.

For a physical asset, insurance value is the cost of replacing a lost asset by an equivalent one, and is thus known as the replacement cost.

If the asset in question is a standard catalogued item, then a replacement cost for a new item will be readily available. Allowance must, however, be made for wear and tear and, unless the policy was issued on a 'new for old' basis, insurance companies usually pay claims at a figure which is insufficient to replace a lost asset by a new equivalent. The relevant fair market is that for second hand goods, in which the scope for judgement is considerably greater than that for new items.

For unique or unusual assets, such as houses, the value for insurance purposes, although described as replacement cost, is actually the cost of reproducing what was there before the loss, which will normally have been less than total. This 'reproduction cost' approach, although cost-based, relies on hypothetical costs, not on historic costs, and therefore falls into the market category of techniques.

An alternative market approach is based not on replacing a lost asset but on disposing of one no longer required. Again the transaction is a hypothetical one based, in this case, on the orderly liquidation of a business or part of it. Allowance must be made for the costs of carrying out the transaction, giving a 'net realisable value'.

The market basis of valuation is relevant in cases where bankers take assets as security for loans, but in such cases they will tend to be valued on a 'fire sale' or forced sale basis. The value in such cases is substantially lower for most assets than in the case of an orderly liquidation. Much depends on the nature of the asset: a highly-specialised machine tool might be worth little more than scrap value in a forced sale, whereas, given time, it might find a buyer who would be willing to pay a premium price for it; the managing director's car could be sold readily at almost any time.

These market-based techniques allow for inflation. In times of high inflation, accountants and businessmen become more than usually concerned about the artificial nature of the figures appearing on balance sheets, and particularly about the fact that taxation levied on profits after historical cost depreciation is effectively partly paid out of capital. Higher depreciation charges are needed if the value of the assets is to be preserved, and replacement cost is often advocated as the basis for realistic depreciation charges.

As we have seen in Chapter 2, the issue of inflation accounting has never been fully resolved in the UK. On each occasion that it has become current, inflation has moderated before agreement has been reached on the difficult questions involved. The main problem has been that government has been reluctant to reduce its tax base by allowing increased depreciation charges. Without a tax advantage, companies themselves have not been keen to reduce their published profits by making replacement cost depreciation provisions. Some companies which routinely did so, such as Pilkington and Philips, have abandoned the practice to maintain parity of presentation between their accounts and those of others quoted on the stock market.

Income basis of valuation

The third basis for the valuation of an asset is the income which it is expected to produce. This is essentially its value as an investment.

The value is derived as the net present value of the future income stream from the asset, discounted at the appropriate rate. Once the income and the discount rate have been defined, the calculation is straightforward. This value, plus any residual value on disposal of an asset, is also known as its 'value in use'.

The method is obviously most suitable for application to cases in which the income stream is reasonably well defined as, for example, from a rental on property or a regular fee for a licence of some kind. The cost basis is clearly inappropriate in such cases, as the cost bears no necessary relationship to the income derivable from the asset. If a market exists, the market value in that particular use will be much the same as the investment value because that will be the basis of the market valuation. The market may, however, also consider other uses.

The housing market is a good example. For many years past, a combination of rent control and tax and other advantages for home ownership has resulted in the value of a home being substantially higher when owner-occupied than when rented to a tenant. Few new homes were built for rent by private landlords, and tenanted properties which became vacant were normally sold off for owner occupation. Relaxations in the protection of tenants and the changes mentioned above which have made owner occupation less attractive have resulted in an increase in the number of properties being offered for rent.

The other element in the calculation, the discount rate to be applied, is more of a matter of judgement than the income stream, largely because if the income stream were a matter of great uncertainty, other valuation methods would probably be more suitable. The appropriate discount rate has to allow for inherent risk and for inflation, both matters for significant exercise of judgement.

Inherent risk and inflation are closely linked, and there is a considerable danger of using an inappropriate discount rate if this relationship is not properly understood. The starting point for any assessment of the appropriate discount rate is the rate available on government bonds, which are regarded as risk-free. This rate for normal fixed-rate bonds does, however, take inflation implicitly into account. It is therefore superfluous to make an additional allowance for inflation. For those who prefer to make their own assessment of likely inflation, the rate available on index-linked bonds provides a starting point. Further, there may be clauses in the agreement giving rise to the income stream being valued which allow that income to be adjusted in some way to reflect inflation. If so, it should be discounted at a rate which excludes inflation, at least so far as the income is protected from it.

VALUATION OF BRANDS

There is no widely-accepted definition of a brand which, like the proverbial elephant, is more easy to recognise than to describe. The basis of a brand is normally a brand name, which will probably be a registered trademark, and therefore a recognisable intangible asset in its own right, but the concept of a brand also incudes all the other features which make it distinctive and valuable. These features will vary with the product or service, ranging from the secret recipe for Coca Cola, the combination of ingredients and the shape of the still which combine to form the flavour of a Scotch malt whisky, through Marks and Spencer's reputation for changing any goods without demur, to the image of power, glamour, or dependability which advertising confers on different makes of consumer goods or services.

An occasion for valuing a brand obviously arises when it is sold or acquired, either because the buyer and seller have to agree on a figure in a stand-alone transaction, or because a company which has acquired one or more brands as a result of a takeover of another company may wish to put a value on it or them for balance sheet purposes, as it is allowed to do, since the acquisition was 'for valuable consideration' (see Chapter 2).

Chapter 2 showed how interest in capitalising as much as possible of the intangible assets acquired in a takeover increased once this component of the purchase price became significant in the early to mid 1980s, when a quirk of financial accounting practice made this course of action attractive to managements. In a takeover, the value of brands is one of the components of the goodwill element which can either be put on the balance sheet as an asset and depreciated over a period of years, or else written off at once against the reserves of the company. The financial position of the company is entirely unaffected by the accounting treatment, but capitalisation and amortisation has an effect on future profits, since the amortisation appears as a charge on the profit and loss account, whereas immediate write-off affects the balance sheet only, effectively reducing the amount accumulated from past profits.

A brand is valuable because the goods or services sold under its banner either command higher prices or sell in greater volume, or both, than competing goods or services of similar quality. The excess profit arising from the combination of extra margin and volume is thus the starting point for the valuation of a brand. Although most attention has been paid to the controversial issue of putting brands on the balance sheet, as discussed in Chapter 2 above, in the long run the more important application, certainly for the purposes of this book, is in connection with management accounting where the value of brands and other intangible or intellectual assets, and particularly whether or not that value is being maintained or enhanced, is a major consideration.

Interbrand methodology

When companies began to put brand values on their balance sheets in the late 1980s, they gave no details of how the values had been derived. The effect of the change was, however, so significant in some cases that considerable interest was aroused both in the desirability of the practice and the details of how it had been implemented. Ranks Hovis McDougall (RHM), which was the first company to include in the figure on its balance sheet the value of brands developed within the group as well as for those acquired from outside, increased its net worth from £301 million without brands to £979 million with them. It therefore bowed to the pressure from investors and others by publishing some details of the methodology which it had used to derive what it described in its statement of accounting polices as a 'current cost', and which had been devised by the consultancy firm Interbrand. The methodology has since been described by Penrose (1989).

The RHM/Interbrand approach started from the controversial point of view that the balance sheet purports to show the underlying financial strength of the business. Accountants and analysts would both reject this view, the former on the grounds that the balance sheet is designed to show the ascertainable financial position of the company, with its particular contribution being to show the sum of costs which have gone into tangible assets and therefore not yet charged against revenue; the latter on the grounds that it is cash flow which matters, and all intangible assets are either irrelevant or have values which are derived from estimated cash flows by judgements which they would prefer to make for themselves rather than have them made by managements.

RHM and Interbrand rejected all conventional methods of valuation in the application to brands for the following reasons:

(a) Valuation of a brand by reference to the cost of its development was clearly inappropriate, since the development might have been abortive or indifferently successful. It is results which count, not the cost of getting them.

(b) Market value was unlikely to be available and, in any case, since brands are not developed in order to be traded, inappropriate. Further, market-based values for intangibles are not allowable for balance sheet purposes under the Companies Act.

(c) Valuations based on discounted estimated future cash flows are difficult and uncertain, and the results unlikely to qualify as acceptable for balance sheet purposes.

(d) Many brands cannot be said to command a price premium relative to generic competitors because such competition does not exist. The value of a brand depends more on the assurance of continuity of demand which it provides than on any price premium.

Methods based on consumer-related factors such as esteem or awareness were also rejected on the grounds that they would bear no relation to commercial reality.

The alternative methodology proposed was to take the past profits attributable to the brand and apply to that figure an earnings multiplier representing the brand strength to arrive at a figure for the brand value. The steps in the valuation were:

1 Past profits were adjusted to allow for factors unconnected with the brand itself, and an average over several years – three in the RHM case – taken to allow for year-to-year fluctuations.
2 The brand strength was estimated by reference to 'leadership' (its position in the 'pecking order' in its market); 'stability' (long establishment of customer loyalty); market (reflecting the greater volatility of markets, for example fashion goods in relation to foods); internationality; trend (defined, rather vaguely, as the ability to remain contemporary and relevant to consumers); support (in terms of quality and quantity of advertising, etc); protection (legal, either through trademarks or other rights). Each of these aspects was scored in what was claimed to be a consistent, logical, and verifiable manner, and the total score expressed as a percentage.
3 The earnings multiplier was determined from the brand strength using an S-shaped curve. The highest possible value of the multiplier is less than that for a risk-free investment, to allow for risk, and may vary from business to business and industry to industry. The range of multiples is from above to below the average P/E ratio for the relevant industry.

It is claimed that this methodology, by considering the marketing, financial, and legal aspects of brands provides a reliable and consistent method of valuation for balance sheet and internal management purposes. The methodology certainly reflects the experience that in well-managed and successful companies the intellectual capital increases with time, along the lines referred to by TI (Chapter 2). Intellectual capital also tends to disappear from failing companies before more current measures of performance, particularly profit, are affected.

Relief from royalty basis

A more orthodox approach to valuing brands, which can also apply to other intangible assets, is the 'relief from royalty' approach. The concept here is that if the owner of the brand or other asset did not, in fact, own it but licensed it from the owner, then a royalty would be payable. This virtual royalty is therefore a benefit of ownership and can be valued.

OPTIONS AND THEIR VALUATION

As the financial markets have become more sophisticated they have made increasing use of derivatives which are secondary forms of property. The two basic types are futures, which are contracts to buy or sell some form of property at some future time at a specified price; and options, which are renounceable rights to buy (a 'call' option) or sell (a 'put' option) at a specified price (the exercise price) on, or sometimes at any time before, a specified date.

Futures enable the participants to remove the risk that the market for something which they expect to buy or sell at some point in the future may move against them in the interim: options enable them to insure against the risk that their expectations are disappointed.

Derivatives have long been used in commodity markets whereby a farmer, for example, may sell his growing crop forward in order to ensure an income until the harvest, possibly insuring himself with an option in case his own crop fails totally or partially. The counterparty, the purchaser, for example a food manufacturer, assures itself of the raw materials which it needs, and may insure itself with an option to buy. Foreign exchange futures and options similarly enable those trading overseas to remove the risk that exchange rates may move against them by buying or selling foreign currency in advance of an expected transaction. Again, offsetting transactions can be entered into as a form of insurance.

In common with other markets for assets which are standard and easily transportable, the markets in derivatives have taken on lives of their own with most business being done between traders rather than between traders and ultimate users. The range of assets on which derivatives are traded has also increased, with those on shares being a particularly active market. Options on shares are of interest in the context of innovation for, as has been mentioned above, one way to look at R and D projects is as providing options for carrying on further into exploitation and sales. The implications of this concept will be explored below, but we first describe the method which has been developed for valuing options on shares.

As with any valuation, the object is to decide whether the returns will justify the cost or, putting the problem the other way round, what price should be paid to obtain the returns envisaged. The price will be fixed but the returns are uncertain, so the problem is to assess what the chances are of the return exceeding a particular value.

One major input to the calculation for stock options is the well-established observation (which originally came as a shock to the statistician who discovered it), that share prices move randomly. They follow a random walk pattern of movement, as in Brownian motion or diffusion processes in general. Thus, the greater the interval between the purchase of the option and the time at which it is exercised, the greater will be the range of possible prices. Different shares have their own price diffusion rates.

Further, as investors expect a constant, but different, return on each share whatever its price, the price will also drift to reflect that return at a rate which is constant as a proportion of the price. The return is conventionally given the symbol μ, so that the drift rate is μS, where S is the share price, or:

$$dS = \mu S.dt, \text{ which gives:}$$

$$S = S_0.e^{\mu t}$$

The combined result of the drift and the random variation is that the distribution of share price over time is lognormal. The log part comes from the drift, with the normal part being the result of the random walk or diffusive behaviour.

The normal distribution is, of course, very well understood and characterised, so that the chances of the share price being above the exercise price (the price at which the option can be exercised) can be calculated, given the drift rate, which represents the expected return on the share, and its dispersion, which represents the sensitivity of its price to market conditions. Both of these can be derived from observations of the price of the share under consideration.

We still need one more piece of information before we can decide what the option is worth. This is the odds at which we are prepared to take the bet that the option represents.

The key to resolving this conundrum was provided in 1973 by Black and Scholes, who pointed out that it was possible to set up a risk-free equivalent to the option. Its value must therefore be the value of the option.

The risk-free portfolio depends on the fact that the value of the share and the value of an option on it are affected in the same way, but to different degrees, by changes in the market. If the share price goes up, an option to buy it at the fixed exercise price at a future time also increases in value. If the share price goes down, so will the value of the option. Following somewhat similar lines to those open to the farmer and the businessman above, the risk-free portfolio for an option to buy consists of selling the option and buying sufficient shares to ensure that any change in the value of the option will be counterbalanced by the corresponding movement in the share price. Since this portfolio is risk-free, the return on it must be the going rate for risk-free investments. If it were not so, the market would correct the prices until it was. The conditions apply only at a specific time, with the ratio of shares to options in the portfolio varying with their prices. Symbolically, the value of the portfolio is:

$$-f + (\partial f/\partial S)S$$

where: f is the value of the option
S is the value of the share

Black and Scholes incorporated this portfolio into the equations describing the random walk or diffusional behaviour of share prices, and finally arrived at a differential equation which expresses the relationships between the various factors:

$$(\partial f/\partial t) + rS(\partial f/\partial S) + 1/2\sigma^2 S^2(\partial^2 f/\partial S^2) = rf$$

where: t is time
r is the risk-free rate of interest
σ is the standard deviation of the proportional change in the share price

It can be seen that the drift rate, which appears in the equation describing the variation with time of the share price, disappears from the Black-Scholes equation as a result of its equivalent effect on the share and option prices.

Differential equations of this type admit of many solutions, depending on the boundary conditions, and they have been extensively studied. Black and Scholes solved their equation for the specific case of 'European' options, in which the exercise date is fixed ('American' options allow exercise before the due date, complicating the mathematics even further), arriving at a formula which incorporates cumulative probability distribution functions for a standardised normal variable.

Values for this function have been tabulated and are available, for example, in Brealey and Myers (1991) or Hull (1993), and in some computer packages aimed at the financial market.

The determining factors for the option price are the ratio of the share price to the present value of the exercise price and the product of the standard deviation of the share price and the square root of the time until the option is exercised.

The value of the option increases, as common sense would suggest, with the ratio of share price to the present value of the exercise price; and also, less intuitively obviously, with the product of standard deviation and time. The interpretation of this latter dependence is that, as the range of possible prices increases with time and the inherent variability of the particular share price, so the chances of it having a higher price than the exercise price increase. The wider the range of possible share prices at the exercise date, the greater the chance that the eventual price on that date will be above the (fixed) exercise price. Options on risky shares are worth more than equivalent options on safer shares.

The ratio of the rates of change of price for the option and the share, $\partial f/\partial S$, which is known as the 'hedge ratio', is conventionally given the symbol δ (delta). It depends on the same factors as the value of the option – the ratio of share price to the present value of the exercise price, and the product of standard deviation and time. Tables for finding 'option deltas' are available in the same way as tables for valuing options.

Empirical evidence seems to be that the Black-Scholes equation adequately represents option prices, and it is in widespread use. Further development of pricing models for options and other derivatives still continues, providing employment in the financial world for mathematicians and theoretical physicists at salaries which they would be hard-pressed to obtain in scientific occupations.

As an illustration, or in cases where only a few estimates of possible outcomes are available, the concept of the equivalent risk-free portfolio can be used directly. Instead of buying the option we could buy the number of shares determined by the hedge ratio and borrow a certain amount of money at the risk-free rate. When the option matures we can repay the loan, plus interest, by selling some of the shares. In the worst case we would have to sell all the shares and would finish with nothing, a situation exactly the same as buying an option with an exercise price at or below the price of the shares on the maturity date. Alternatively, if the share price has increased since we bought them sufficient to pay off the loan, plus interest, and leave a profit, we shall be in the same situation as if we had bought the option.

Formally, this approach is equivalent to rearranging the equation for the value of the Black-Scholes portfolio, L:

$$L = -f + \delta S$$

as:

$$f = \delta S - L$$

We can use estimates of the position at the maturity date of the option to calculate both δ and the amount of money, L, to borrow. A simple example, using only maximum and minimum estimates for the value at the maturity date, $t = 1$, is shown in Table 3.1.

Table 3.1 Valuation of option

| | Value of investment | | |
| | t=0 | t=1 | |
		Down	Up
Buy option	f_0	0	f_u
Buy shares, take out loan	$\delta S_0 - L$	$\delta S_d - L(1+r)$	$\delta S_u - L(1+r)$

Source: DWB

As the two possibilities are equivalent, by definition, we get:

$$\delta S_u - L(1+r) = f_u$$

and:

$$\delta S_d - L(1+r) = 0$$

hence:

$$f_u = \delta(S_u - S_d)$$

from which:

$$\delta = f_u/(S_u - S_d)$$

and:

$$L = \delta S_d/(1+r)$$

We would know the share price, S_0 at the time of decision (t=0); the exercise price, S_e; and the risk-free rate of interest, r. Using our estimated values for S_u and S_d, we can calculate:

$$f_u = S_u - S_e$$

and, finally, the value of the option:

$$f_0 = \delta S_0 - L$$

For example, suppose we have an option to buy shares at a price of 110 in one year's time. The current price is 100, and we estimate that the chances are 50 per cent that the price will rise to 130 and 50 per cent that it will fall to 90. If the current risk-free interest rate is 7 per cent, then the calculation is:

$$\delta = (130 - 110)/(130 - 90) = 0.5$$

$$L = (90 \times 0.5)/(1.07) = 41.47$$

$$f_0 = (100/2) - 41.47 = 8.53$$

Or, in words, the value of the option is 8.53, and the risk-free portfolio would consist of half a share and a loan of 41.47.

APPLICATION OF OPTION VALUATION TO R AND D

Once the Black-Scholes solution was available, it was not long before its application was extended to other types of option for investment, such as mineral rights. R and D projects are another case in which there has been interest in using the technique, on the argument that an R and D project, if successful, essentially provides an option to make the subsequent investment in marketing, production, and so on which are required to exploit the R and D results in order to produce an innovation. If the likely returns from the investment are unattractive, then it will not be made: the cost of the R and D project is, however, not wasted, but can be regarded as the price paid for opening up the opportunity of making the investment. Estimates of the returns on the innovation are likely to be much more accurate after the R and D stage than before it, and it is therefore unwise to use preliminary estimates of return on the innovation as the sole foundation for the decision on whether or not to embark on the R and D (Newton and Pearson 1994).

The same argument can be used for an otherwise unattractive product innovation or process plant investment if it opens up opportunities for Mark II and successor versions which might have better commercial prospects. As a strategic argument, this line of thinking has much to commend it. R and D is, by definition, a process with an uncertain outcome, and the first version of a product is unlikely to be the definitive one. Predictions of what opportunities could be opened up can never be very precise because of the great uncertainties about the exact performance, the scope for further development, and the competing products and other market conditions which will obtain by the time any product could be offered. All these things will be clearer once the R and D has been done, the Mark I product made, or the pilot plant installed. The first attempt is likely to be more useful for what it teaches than for the returns it brings, but the costs of waiting and making a later entry to the market may be much higher.

However valid the strategic argument may be, the applicability of the precise valuation technique derived for financial instruments depends on how far its assumptions correspond to those applying to the particular project, and how sensitive the results are to changes in those assumptions. A comparison for R and D projects is shown in Table 3.2.

Table 3.2 Comparison of financial options and R and D

Financial options	R and D case
Option price (fixed)	R and D project cost (estimated)
Exercise price (fixed)	Innovation investment (estimated)
Share price (normally distributed)	NPV of innovation (variable)

Source: DWB

It can be seen that the critical features of asymmetry and uncertainty about the value of the eventual asset to be purchased are common, but that the values of the various factors in the R and D case are less well defined than those in the options case. In particular, the cost of capital used to calculate the net present value of the eventual innovation must be that which reflects the cost of capital to the company, not the risk-free rate. This subject is explored further below.

All the factors in the Black-Scholes solution would be known for an R and D project except the standard deviation of the NPV of the innovation. Even so, there will be some estimates of the range of possible returns from which a standard deviation can be calculated, and there may well be a wealth of previous experience of projects which can be used to derive a more reliable figure.

Table 3.3 gives some illustrative figures for an R and D project which costs 100 units, after which an investment of 1,000 units would be needed to make an innovation. Delays of two and five years are taken for high risk and low risk investment projects.

Table 3.3 Value of R and D Options

Return	2 years		5 years	
	Low risk	High risk	Low risk	High risk
500	0	30	5	115
1000	140	380	290	460
1500	600	660	780	885

Source: DWB
Note: R and D cost 100; follow-on investment 1000, discounted at 7 per cent; return is NPV; low risk has standard deviation 0.1, high risk has standard deviation 0.4

The effects of risk and time show up most clearly for the project whose returns, expressed as net present value, are equal to the investment cost (at the time of the investment). The low return project is unattractive, regardless of risk or the time before it would start; whereas the risk and time have little effect on the attractiveness of the high return project.

Questions for discussion

1 How reliable are capital values based on estimated cash flows?
2 Under what circumstances would the use of formal option valuation techniques be useful for making R and D decisions? Would Pilkington's development of float glass have been helped by viewing it as an option process?
3 Could the brand valuation methodology be applied to a company's reputation as an innovator?

Chapter 4

Innovation and management accounting

MANAGEMENT ACCOUNTING

The role of management accounting

Management accounting is concerned with providing information for planning, decision-making, and control of business enterprises. It is the aspect of accounting which impinges most directly on most employees and managers in companies but, because of its focus on the internal workings of the company, it is less standardised, less well documented, and more flexible than financial accounting. In the context of innovation, while financial accounting may be criticised for taking less notice of provisions for the future than even its necessarily backwards-looking orientation excuses, management accounting may be criticised for being so concerned about the future and the present that it pays little heed to the lessons of the past.

In this chapter, we will look briefly at some of the main concepts in management accounting, and then go on to discuss in more detail the aspects and developments which are of relevance to innovation. Much of management accounting is concerned with matters of a more predictable nature than innovation, and readers who are interested in these aspects should consult specialist textbooks on the subject.

Management accounting arose in response to needs which became evident in some cases several hundred years ago, but its major development occurred from the mid-nineteenth century when the scale of industrial enterprises such as textile mills and railways became sufficiently large for the managers to be unable to rely on less formal methods of gathering information on which to base decisions.

During the early part of the present century, management accounting became progressively more the province of specialist accountants rather than of the engineers and managers who had given birth to it. It became normal for management accounting and financial accounting to be parts of the same company-wide accounting system in order to ensure consistency and minimise the amount of effort devoted to collecting and analysing figures. While these were obviously good reasons for the development, it has been powerfully argued by Johnson and Kaplan (1987) that, as a result, management accounting has become subservient to financial accounting and has in consequence lost its relevance to management.

In particular, Johnson and Kaplan argue that management accounting has not kept pace with technological developments and has, in consequence, become a handicap, rather than a help, in decision-making. There is undoubtedly some truth to this contention but, not least due to the efforts of Johnson and Kaplan themselves, there have also been significant developments in recent years both in the strict field of management accounting and in the wider field of performance measurement to which it has given rise. Aided by the availability of cheap computing power, there are distinct signs that management accounting and performance measurement are once again developing identities separate from financial accounting.

Absorption costing

As an illustration of the need to keep management accounting techniques in line with technological developments, the case of computer-aided manufacture (CAM) is illuminating. The absorption costing method for products, which is widely used, is based on the costs of materials, direct labour, and overheads. The overheads contribution includes all the general costs of the workshop and of the management and support services which cannot easily be allocated to particular products, but which are shared between all the products being made.

The approach ensures that all the costs incurred by the firm are allocated to products, thus keeping the internal management accounts, concerned with costing products, in line with the financial accounts, concerned with total costs. When this method was devised, overheads were a relatively small proportion of total costs and so it made sense to base the costing system on the main contribution, direct labour, to which a relatively small percentage addition was made for overheads in arriving at total costs.

With the development and introduction of automated machinery, however, the relative costs of overheads and labour changed dramatically, with overhead rates rising to levels of 1,000 per cent or even 2,000 per cent in some cases. To carry on using direct labour costs plus overheads as the basis of a costing system in such cases is clearly misleading. First, it focuses attention on what is obviously only a minor contributor to costs, namely direct labour, diverting attention from the more important contributions. Second, it may tempt management into thinking that a large saving can be made in total costs by saving direct labour, since this is what the costing system indicates. What happens in practice, of course, is that any saving in direct labour costs – usually by reducing manning levels – merely puts up the already-high overhead rate on such direct labour as remains, with little effect on total costs.

The direct labour plus overhead basis is nevertheless still a reasonable one to use in costing labour-intensive processes, of which R and D is certainly one. Even here, however, overhead rates are by no means low, with typical industrial figures being in the range 100–150 per cent (see Table 2.12).

Marginal costing

In the context of planning, an alternative way of looking at costs – the marginal costing method – is more helpful than the absorption costing method. Marginal costing can trace its origin to the pioneer of mechanical computing, Charles Babbage, who was the first person to point out, in the early 1830s, that the costs of manufacturing could be divided into those which were fixed, regardless of output, such as the rent of the factory and the cost of running the managing director's car, and those which varied with output. Babbage was also an advocate of using cost accounting as a means of identifying promising areas for cost-reducing innovation (Rosenberg 1994).

The distinction between fixed and variable costs is not quite the same as that between direct costs and overheads, although there are similarities between the two divisions. For example, direct labour costs are to some extent fixed in the short run, rather than variable, since people cannot be laid off and taken on again at short notice, and there may in any case be agreements for minimum pay or even fixed pay for a period. R and D staff, for example, would normally be on fixed salaries. Similarly, costs of power will be variable with output, but usually treated as overheads for absorption costing purposes.

The distinction between fixed and variable costs is relevant to the makeup of profit. Until the fixed costs have been covered, a start cannot be made on making a contribution to profit. Assuming that everything made is sold, the total costs of the firm rise with production from the basis of the fixed costs, while the sales revenue rises from zero. If the decisions on costing (based on management accounting) and pricing (based on marketing considerations) have been made appropriately, the sales revenue eventually reaches the total costs at the 'break even' point, and then exceeds them, bringing in a profit (Figure 4.1).

It is assumed in Figure 4.1 (whose figures are arbitrary) that the cost and revenue curves are linear, that is that the cost of producing each item is the same and the revenue from its sale is also the same, regardless of production volume. In practice, the cost is likely to decrease with volume, for reasons which will be discussed below, while the revenue may increase or decrease depending on the market conditions.

It can be seen that profit is a function of the whole operation not of an individual sale. The concept of the profit on a specific sale is valid only in the context of other sales or of assumptions about them. For this reason, in hard times, companies may be willing to sell items at prices which are not 'profitable' under the usual assumptions about cost structures but which, provided they

Cash flow

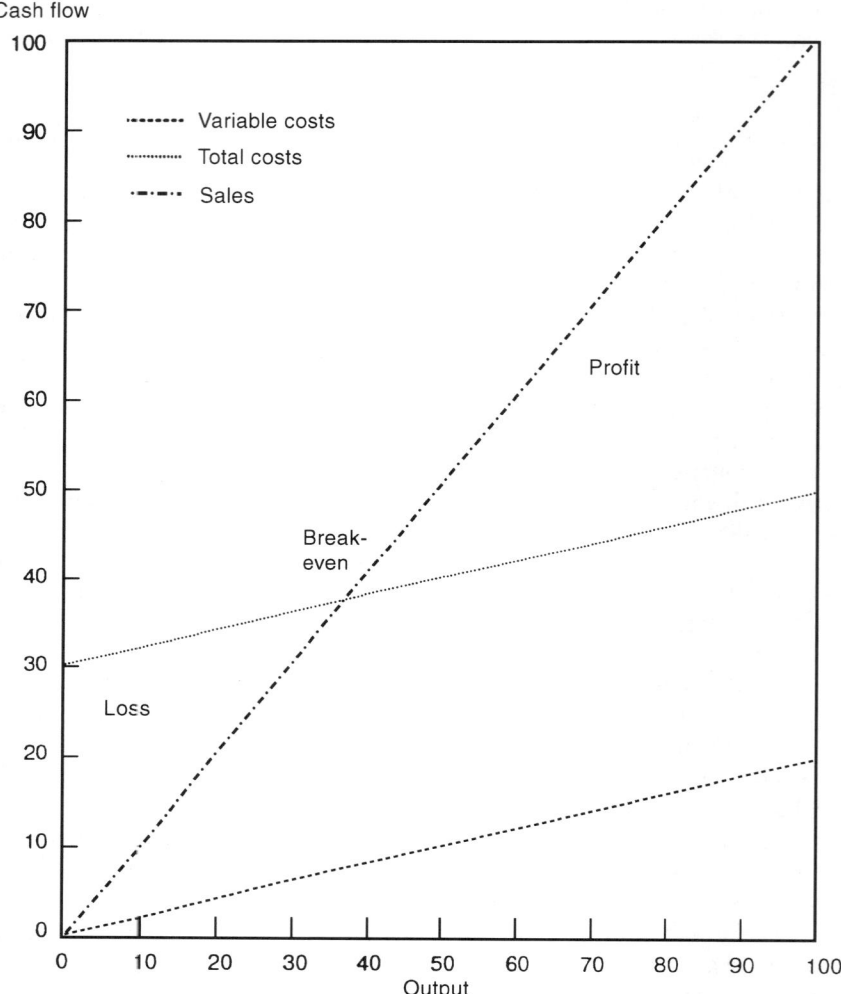

Figure 4.1 Marginal costing
Source: DWB

are higher than the variable costs of production, are worthwhile because they
bring in some revenue to offset fixed costs. Provided this condition is satisfied,
then every sale makes a contribution to profit, with the total contribution
rising with the sales volume.

The concept of 'contribution' is a particularly useful one when, as is almost invariably the case, the firm is making a mix of different products. One product may be unprofitable, but its contribution may nonetheless be impossible to replace, at least in the short run, by increasing the sales of existing products or introducing new ones. In such a case, dropping the unprofitable product would worsen the overall position of the firm rather than improving it. If one factor of production is limited in capacity, and some internal rationing has to be imposed, then contribution related to that scarce factor is a useful guide to optimising the product mix. Comparing different products on the basis of contribution as a proportion of sales has the advantage of avoiding the problems of distortion which may be caused by the allocation of overheads in absorption costing.

Standard costing and budgeting

For planning and control purposes the concept of standard costs is often used. Standard costs are estimated as those which should be achievable for the product or products to be made. They constitute targets for those responsible for the production, against which actual performance can be compared on an ongoing basis. Departures from standard, known as variances, are monitored, and investigation made and action taken when they reach significant levels.

Standard costing is closely related to budgeting which is the process of setting out a plan for the expenditure and income of a firm and the units into which it is divided for management purposes.

Activity-based costing and management

Activity-based costing (ABC) is a relatively new technique which has been developed in response to some of the shortcomings, especially as regards allocation of overheads, of standard absorption costing. It attempts to allocate costs, as its name implies, on the basis of the activities which give rise to them.

The starting point for an ABC system is a careful analysis of the activities or processes which actually take place. A favourite example of the difference which it can make is in the costs of invoicing and order-handling generally. Such costs, which are largely incurred in a service department, are usually allocated, under an absorption costing system, according to sales volume or some similar general measure. In practice, the costs of servicing an order are largely independent of the size of the order – it is certainly no more difficult or expensive to process an order for 1,000 units than one for ten units – and so ABC allocates order-handling costs by the number of orders rather than by their value. Similar changes can take place in the costing of the manufacture of large and small orders if a significant amount of expense, usually in the form of direct labour, is involved in setting up for each order.

The result is that small orders can be shown to be considerably more expensive relative to large orders under an ABC system than they are under a conventional system. This finding, in turn, may have serious implications for company policy. A company which specialises in small orders, and prides itself that they are more profitable on its absorption costing system, may find that on the more realistic basis provided by ABC they are actually less profitable.

Accurate costs are particularly necessary when the mix of products is changing quickly. A costing system may be inaccurate, but adequate, if the product mix remains the same, since it ensures, in an unchanging world, that revenues cover costs and provide a profit.

If it did not, it would be changed or the company would go out of business. If the product mix changes, however, in response to demand, and the costing system is inaccurate, then there may be unexpected effects on profit. If, in the case discussed in the previous paragraph, a company deliberately sets out to increase the proportion of small orders, on the grounds that they are more profitable, it will find, if the costing system is inaccurate, that the result of its efforts will be to damage profit.

The evidence is that ABC has not as yet been widely adopted in the UK, and that it is much more suitable for some circumstances than others. If overheads can indeed be traced more accurately than has normally been done, as in the example given above, then ABC can be helpful. Not all cases are as simple as this, however, and the utility of ABC has by no means been universally accepted.

PROJECT EVALUATION

Selection of which investment project or projects to pursue out of a wide range of possibilities is perhaps the most important application of management accounting in decision-making. A more-or-less certain amount of money has to be invested over a comparatively short time in order to obtain a more-or-less certain return, usually spread over a longer time, and the issue for management is whether the investment is worthwhile at all, and if so, what the returns will be.

The most obvious case is when the investment in question is in new production machinery, where the investment is usually made in one step, and the return comes in the form of profits over the period of its use. The techniques for dealing with such cases have been well developed, although in practice less often used than written about, and they have subsequently been applied to innovation, whose various components are essentially project-based. The basic approach is that of income-based valuation (Chapter 3) in which a forecast is made of the costs and returns from the project on a year-by-year basis, or other suitable interval, to which some form of analysis is then applied in order to determine whether the project is worthwhile, particularly in relation to its rival projects and to the company's cost of capital.

There are four basic methods of assessing the return: payback period, rate of return, internal rate of return, and net present value.

Payback period

The payback method is the simplest, much criticised, but still popular, method of investment appraisal. It simply calculates how long it will be before the initial investment is recovered through the cost savings or extra net receipts which it will produce. Thus, if a machine costs £50,000 and is expected to bring in an annual return of £15,000, then the payback period will be just over three years. This figure would be regarded as too high by many British companies who tend to favour periods of two to three years.

The advantages of the payback method are that it is very simple to use, and gives an intuitive feel for the value of the investment. Its only assumption is that of the return and its timing, which is a necessary foundation for any investment appraisal.

The technical disadvantages of the payback method are that it takes no account of any benefits accruing after the payback period, nor of the time value of money. Over a short period such as two years, and given the uncertainties of the estimates of return, this latter weakness is hardly a severe one in times of low inflation and interest rates. If a return can be achieved in two to three years, then the project is obviously worthwhile, and the method may be satisfactory.

Ignoring the longer-term benefits is, however, a serious weakness if the project is likely to pay off only over a longer term. For example, such projects as the extraction of oil from the North Sea and the construction of the Channel Tunnel, even on its original optimistic cost estimates, would be ruled out from the start by a payback method of appraisal. The same applies to innovation projects in most industries although timescales differ. A typical period for the development of a new drug is, for example, twelve years (ABPI 1993), and figures of around three years are regarded as state-of-the-art for automobile development (Lutz 1994). The returns take at least as long to come in.

Accounting rate of return

The accounting rate of return method takes the estimated total profit over the life of the project and relates it to the investment to get an average rate of return on investment for the total project life.

While this technique is compatible with financial accounting concepts, it is not helpful in comparing projects because it is insensitive to differences of timing of cash flows and is unable to compare projects with different lifetimes.

Discounted cash flow methods

Discounted cash flow (DCF) methods make specific use of the concept of the time value of money, and are theoretically superior to the methods discussed above. They take two main forms, based respectively on internal rate of return and net present value.

The internal rate of return (IRR) method takes the estimated cash flows from the investment over its life, and calculates the discount rate which would produce a net present value of zero. Projects which produce a rate in excess of a chosen 'hurdle rate' are accepted, while those below are rejected.

The IRR method is closely related to the net present value (NPV) method, which is that favoured by management accounting theorists. Again, the NPV method starts from the cash flows estimated to result from the investment, starting with the large negative flow arising from the investment itself, and discounts them to present values at the start of the project, this time using a fixed discount rate chosen to represent the company's cost of capital or some minimum hurdle rate regarded as acceptable.

Projects are acceptable if their NPV is positive, with those with the highest positive values being the most desirable, other things being equal. Unlike the IRR method, the NPV method is sensitive to the absolute size of an investment.

The technical advantages of both the IRR and NPV methods are that they allow for the time value of money and permit comparison of projects with different patterns of cash flow. Their primary disadvantage is that, given that discount rates tend to be of the order of 20 per cent or so, they share with payback the bias towards projects which pay off in the short term and the corresponding bias against longer-term projects.

As with payback, the bias against longer-term projects affects innovation particularly badly, and one prominent industrialist even went so far as to call DCF 'the enemy of strategy' (Malpas 1991). His criticism was aimed at the unthinking use of DCF calculations as definitive criteria for acceptance or rejection of projects rather than, as they should be considered, aids to decision-making.

This disadvantage is particularly important for innovation projects, where the uncertainties of the estimates of future cash flow on the one hand, and the strategic necessity of renewing the business on the other, make the exercise of managerial judgement even more important than in other investment decisions.

A further problem with DCF methods in practice is, as discussed in Chapter 3, that users tend to allow for inflation in the discount rates which they use, since they base them on current experience, but ignore it in making forecasts of future cash flows, for precisely the same reason. This effect produces a further, and entirely unwarranted, bias against longer-term projects.

The major factor not considered in a straightforward DCF calculation is the opportunities for further development which a project might open up, and Malpas later came to the view (Malpas 1991) that his criticism of DCF techniques was not true if 'options for the future' were taken into account. Before turning to this subject, however, we shall look at an intermediate stage in which appraisal techniques were developed for assessing the intangible benefits from investment in advanced manufacturing technology (AMT).

MEASUREMENT OF INTANGIBLE BENEFITS FROM INVESTMENT IN AMT

The original application of computers to manufacturing was for automating processes which had formerly been controlled by skilled workers. The savings in labour costs amply justified the investment which could therefore be assessed adequately by the orthodox techniques discussed above. As the technology developed, however, the benefits began to be felt in such intangible aspects as increased flexibility of production, higher quality (in the sense of consistency of conformance to specification), more rapid response to customers' orders, and improved image in customers' eyes. While all these were worth having, it was difficult to estimate their value in ways which could be incorporated into the financial assessment process. Since the scope for further labour cost savings was small, managements were faced with a situation in which the investment could not be justified on the basis of the figures and yet they believed it to be essential to stay in business.

Some managements, especially where the engineers were dominant, took the view that strategy over-rode financial calculation, and went ahead with the investment anyway. Some adjusted the figures until they got the answer which their financial appraisal system demanded (see, eg Currie 1989). Some, in companies where the financial function was dominant, were guided by the figures and suffered the competitive consequences. The situation was far from satisfactory and academic workers at UMIST therefore developed, in 1985, a method of evaluating these intangible benefits (Primrose 1991).

The approach pursued by Primrose and his associates was based on the recognition that the benefits of the investment would appear, not where the expenditure was made, but elsewhere in the company. While the cost of a new machine of advanced specification would be charged to the production department, the main benefits, brought about by the improvements to intangible factors, would appear elsewhere in the company, through increased orders, higher margins, or in other ways. It was therefore necessary to trace the effects through the company and to identify the benefits wherever they might appear. Since the connection, say, between improved quality and increased orders, is not nearly as straightforward as the connection between increased machining speed and production cost of a given part, the benefits had to be estimated rather than measured. Moreover, the best people to make the estimates were not the production engineers but the sales force.

Once comprehensive information about the expected benefits has been collected, a proposed investment can be evaluated by DCF techniques since all the relevant information is available. Not only can the investment be justified without resort to acts of faith, the quality of the decisions will be improved because acts of faith are not always justified. 'Gut feeling' or 'engineer's instinct' may well be a better guide than financial calculation performed on inaccurate and incomplete data, but it is unlikely to surpass calculation based on careful assessment of all the factors, into which instinct and experience will in any case have had an input.

The approach also highlights the fact that the figures in investment appraisal are usually more precise than they are accurate. This fact emerges clearly when sales people are asked to estimate the effects of some improvement on sales, but it is also present in many figures for predicted production performance and similar, apparently much more tightly-defined, aspects of company performance. Because accounting is expressed in numbers, usually to several significant figures in the purely arithmetical sense, there is a tendency to forget that many of those numbers are estimates, and the figures are not as significant in real terms at they appear to be.

This development also naturally brings in the subject of performance measurement. If one of the benefits of a new investment is expected to be a reduction in delivery time, for example, and the financial justification for the investment includes this factor, then it is highly desirable to measure that this reduction has in fact been achieved.

Perhaps the main lesson, however, and one which is relevant to innovation in its more adventurous forms, is that the project under appraisal must be looked at in the context of the company as a whole.

INTANGIBLE INVESTMENT

In the last section we looked at the evaluation of the intangible benefits of tangible investment in advanced manufacturing technology (AMT). The key to the valuation lay in following the consequences of the investment through the business. We now turn to the subject of the returns on intangible or revenue investments, such as those in R and D, training, and advertising. Again, a key concept is the effect of the investment on the company as a whole, but there is a further difference in that investment in AMT shows up on the balance sheet where it can be valued on the basis of its cost, whereas investment in intangibles does not, except in special circumstances, show up on the balance sheet, and cannot in any case realistically be valued by reference to cost. Nonetheless, there is interest in valuing it, either in the special circumstances in which it can be included on the balance sheet, or for internal management purposes where it can help to answer the important strategic management question: are we maintaining the real value of the company in terms of its earning power into the indefinite future?

The importance of measuring returns on revenue investment

Of the main categories of revenue investment or discretionary expenditure identified in ASB (1993) (Chapter 2), R and D is by far the best documented in published accounts, for reasons explained in Chapter 2. Maintenance expenditure is incorporated in costs of sales or other costs and never mentioned in published accounts. Training and advertising expenditure get occasional mentions in the Directors' Report or other text, for those companies in which they are significant, and even rarer mentions in the financial statements themselves (see the British Aerospace example in Table 2.8), but there is no systematic treatment of them. Indeed, when a Select Committee of the House of Commons recommended that investment in training be disclosed in company accounts (HC 1988) in order to make boards and shareholders more aware of the expenditure in this area, the government rejected the suggestion on the grounds that as there was neither a definition nor a specified accounting treatment for such expenditure, such a requirement would be burdensome (DTI 1989).

The Commons Committee explicitly drew the analogy between expenditure/ investment on training and that on R and D, pointing out that the government had then recently expressed support of disclosure of the R and D figure. It implicitly assumed that once boards and shareholders had had their attention drawn to this expenditure they would be anxious to increase it. An earlier survey had, however, shown that training managers were in some cases keen to conceal from their boards the amount spent on training, fearing that once the figure became known there would be pressure to reduce it (Coopers and Lybrand Associates 1985).

Despite the opinion of politicians and other outside observers, company managements require rather more in the way of justification for expenditure than publication of figures which can be interpreted as showing that they are spending less than their competitors, particularly those overseas. If the expenditure is purely a cost, it is obviously desirable to minimise it, but even if the expenditure can be fairly and realistically regarded as an investment, there is still no point in spending more than is optimum for the purposes of the business. Investment is not good of itself, only for the benefits it brings. Since its purpose is to generate future returns, the main tool for estimating optimum expenditure is the estimation of future returns. These can be converted to a capital figure by the methods discussed above if there is some reason for doing so.

Measuring the returns on training and marketing expenditure

Although there have been several attempts to measure the value of training, these have largely been confined to assessing whether it achieved its stated objectives, rather than at quantification in financial terms.

The Interbrand methodology for valuing brands, described above in Chapter 3, can be used to evaluate a proposed or completed marketing campaign. The estimated or achieved impact of the campaign on the brand strength score can be converted to a capital value which can be compared with the expenditure in order to derive a measure of success. Straightforward DCF calculation can, of course, also be used, and there is no reason why both techniques of assessment should not be used, possibly for different purposes. An advertising department, for example, might well be much more strongly motivated to achieve a target brand strength score, which relates directly to its own activities, than to achieving a particular sales or profit target, which would depend on the contribution of other functions.

Measuring the returns on R and D expenditure

The contribution of other functions is essential if any return is to be obtained from R and D expenditure, and it is primarily for this reason that, despite a good deal of activity on measuring the returns on R and D, no definitive, or even widely-accepted, methodology has been developed.

By definition, the R and D phase of a project is over before, probably well before, the stage of obtaining a return through sales or increase in efficiency is reached. The question: what are the returns on R and D? is thus not a very helpful one to ask from the overall company or business point of view, although it may be an important one in the context of internal management, especially that of the R and D activity itself.

There seem to be three possible approaches to evaluating the contribution of R and D to the business. The first is to recognise that R and D opens up possibilities for exploitation which would probably not have existed without it. The opportunities can be valued using the techniques developed for valuing financial options which have been described in Chapter 3.

The second possibility is to recognise that the return comes from the total innovation process, to measure both the expenditure and the returns on that whole process, and to apply DCF or other techniques for evaluating the result.

The third approach is something of a combination of the other two and is to measure the returns on innovations which the R and D has made possible and to make a qualitative judgement as to the relation between the two.

Use of option valuation techniques for R and D

The option approach is especially helpful when the risk of a project is high but, at least in its more precise form of the Black-Scholes approach, it demands a sound statistical base of knowledge about the dimensions of that risk if it is to be used. These conditions are fulfilled in the larger firms in the pharmaceutical industry, since the risk of pharmaceutical development projects is high and the database of projects is large. An informed estimate (ABPI 1993) is that of all the compounds investigated only one out of 5,000 eventually reaches the prescription market. The US-based pharmaceutical company Merck, which was the largest pharmaceutical company in the world until the merger of Glaxo and Wellcome in May 1995, is reported to use option valuation as an integral part of a sophisticated financial modelling approach to its R and D programme (Nichols 1994).

Measuring the returns on innovation

The approach based on the whole innovation process is to measure the total expenditure and the total return on an innovation over its whole life. It is well established that the cumulative cash curve for an innovation is of the shape shown in Figure 4.2. The curve starts at zero and goes down at a gradually-

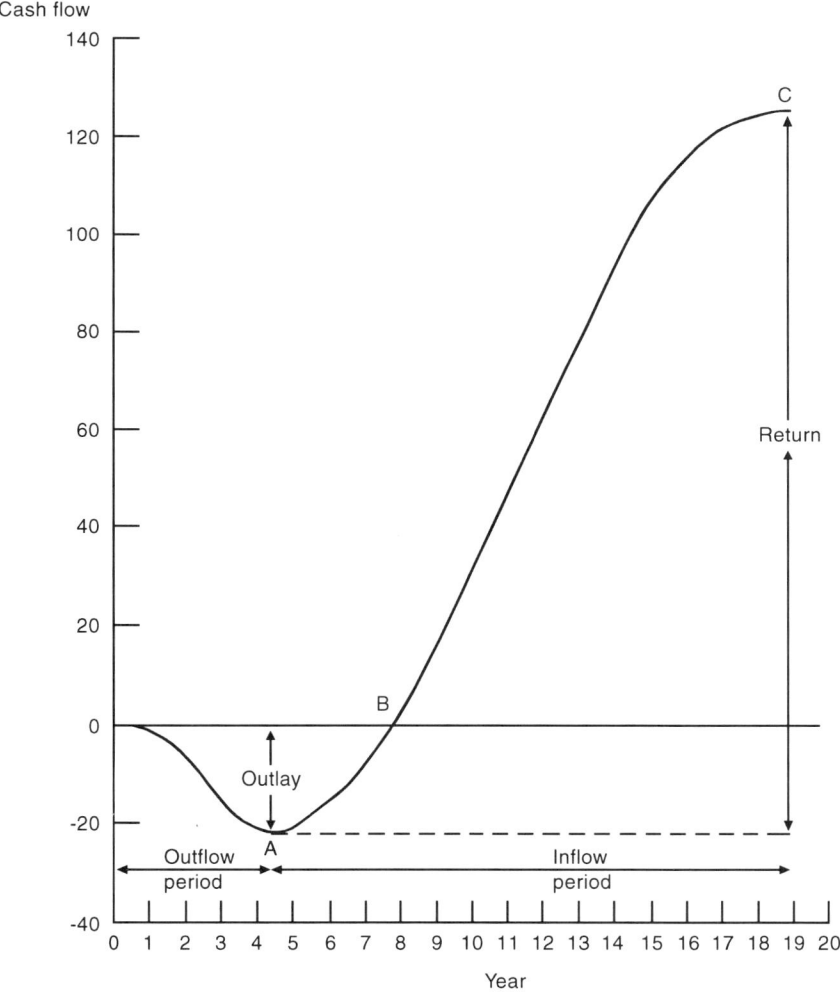

Figure 4.2 Cumulative cash flow for an innovation
Source: DWB

increasing rate as the project proceeds until at some stage exploitation begins, cash begins to come in, the rate of outflow declines, and at point A the cash flow curve turns upwards. It crosses the axis at the point B at which stage all the initial outlay has been recovered and the project has reached break-even. The curve continues upwards in a fairly linear way at first, gradually becoming less steep in its climb as the project begins to lose its appeal or effect, until at point C there are no further returns and the project has run its life.

Although the form of the cumulative cash flow curve for an innovation, and indeed for other types of investment as well, is much the same for any innovation, the exact shape and scale vary considerably between industries. There are three important features:

1 The ratio of outlay to return, which determines how much the company needs to spend on innovation.
2 The ratio of development time (cash outflow period) to market life (cash inflow period), which is a major influence on the number of developments which need to be pursued at any one time.
3 The absolute sizes of both the spend and the times involved, which are major strategic factors whose significance will be explored in more detail in Chapter 6 below.

Discounted cash flow assessment of innovation

The cash flow of Figure 4.2 has an outflow of 21 units and a return of 146 units, giving an undiscounted net balance of 125 units. The annual cash flows can be discounted to give a net present value for the project at the time of its inception by the technique described above. The NPV calculations are given in detail in Table 4.1 for a discount rate of 10 per cent, at which it can be seen that the NPV of the project is 32.64. The NPV for the same cash flow figures is 15.88 at a discount rate of 15 per cent and 6.67 at 20 per cent.

Other uses of the innovation cash curve

The cash flow curve for an individual innovation is a useful conceptual tool but its value for assessing the worth of a proposed project depends on the accuracy with which the figures can be estimated in advance. Such estimates are much helped by knowledge of past projects, and the collection of figures for projects under way can thus have two benefits: it enables the progress of the current project to be monitored, and it helps to build up a database of past experience which can be used when making choices amongst possible future projects.

The collection of figures for monitoring purposes and their use as an aid to decision-making on future projects receive a good deal less attention in the literature than do the techniques for making assessments in advance (Twiss

Table 4.1 Net present value calculation for an innovation

Year	Discount factor	Cash flow in year	Cash flow discounted
0	1.00		
1	.91	-1	-.91
2	.83	-5	-4.13
3	.75	-10	-7.51
4	.68	-5	-3.42
5	.62	0	.00
6	.56	5	2.82
7	.51	7	3.59
8	.47	10	4.67
9	.42	13	5.51
10	.39	15	5.78
11	.35	15	5.26
12	.32	16	5.10
13	.29	15	4.34
14	.26	15	3.95
15	.24	14	3.35
16	.22	10	2.18
17	.20	6	1.19
18	.18	3	.54
19	.16	2	.33
20	.15	0	.00
NPV			32.64

Source: DWB
Note: Discount rate 10 per cent

1992). This imbalance seems to reflect practice, since a report on evaluation of R and D projects in major European companies (EIRMA 1995) found that only a minority paid attention to post-evaluation of projects, although the amount of effort put into pre-evaluation and the number of techniques employed was large.

One company which has given details of its work in this area is the US multinational Hewlett Packard (House and Price 1991) which claims to have over 50 per cent of its sales in any one year arising from products introduced in the previous three years and to have over 500 product development projects current at any one time.

The first claim is borne out by the brief details of the breakdown of orders by year of product introduction which Hewlett Packard gives in its annual report. In 1993, about 30 per cent of orders were for products introduced in the current year, about 35 per cent for those introduced in the previous year, about 15 per cent for products introduced in the year before that, and around 5 per cent for those introduced three years before, with the balance of around 15 per cent being for products of four or more years' standing. R and D expenditure amounted to 8.7 per cent of net revenue (turnover) in 1993, a drop from the values of 9.9 per cent in 1992 and 10.1 per cent in 1991, although rises in sales meant that the absolute amount spent on R and D increased.

According to House and Price (1991), Hewlett Packard has, since 1987, been using measures based on the cash flow curve for innovations to plan and monitor product developments. The 'return map' as the company calls it, is comprehensible to all the members of a product development team, and so helps to channel effort and attention towards a common goal. The main measures derived from the curve are the break-even time, (at point B in Figure 4.2), the time to market, the break-even after release, and the return factor. The break-even time, at which point the product profits equal the development costs, is regarded as the key measure as it is linked directly with the financial return. The time to market (from the beginning of development) is a measure of development efficiency; the break-even after release (to the market) measures the efficiency of transfer from development to production and thence to the market; and the return factor, which is the total return on the project at a point in time as a proportion of the development costs, gives a measure of overall return up to that point.

The Hewlett Packard case makes an interesting contrast to that of Merck. In pharmaceuticals, Merck's estimate (Nichols 1994) of the cost of developing a drug is the suspiciously precise one of $359 million, a somewhat higher figure than those given by other authorities, although not dramatically so, and the time to market is of the order of ten years. For Hewlett Packard, the risks of each project are lower than in drugs, the development time is typically around a year and the break-even time around four to five years. Given that Hewlett Packard's R and D budget is around $1,500 million per year, the average project development cost must be around $3 million, or less than 1 per cent of that at Merck. We will return to this subject in Chapter 6 below.

Measuring returns made possible by R and D

This approach, which is equivalent to measuring the returns on options which have been exercised, is a simplified version of the previous one, and has been that generally favoured in the past. Attempts to apply it have, however, tended to founder for several reasons:

1 Arguments have arisen about the relation between past R and D and current exploitation, and about the contributions of parts of the company other than R and D.
2 As a result, the systems have tended to become increasingly complicated and doubts have then arisen about their cost effectiveness.
3 As the systems have often been the result of individual enthusiasm, they have tended not to become embedded in the company culture and to be abandoned when personalities or priorities have changed.

The glassmaking firm Pilkington went through a cycle of this kind between one and two decades ago, but in 1992–93 instituted a new system which is reported to be working well, having learned the lessons of the past (Ellis 1994, Houlder 1995). A significant factor in the new system was that the stimulus came not from the R and D function but from a newly-appointed Chief Executive who insisted on having some numerical support to help him justify the company's R and D expenditure to investment analysts.

Bearing in mind this broad requirement, and the fact that its R and D expenditure was essentially constant from year to year, Pilkington decided that it could get a reasonable answer by ignoring time delays between R and D and the returns from its exploitation. It therefore decided to compare current R and D expenditure with current returns from activities which had been made possible by previous R and D expenditure. Having once established this approach as being, it its own words, 'roughly right, rather than precisely wrong', principles were laid down as follows:

> Only major benefits would be taken into account. A law of diminishing returns was likely to set in if small projects were included.
>
> The measurement of benefit would be based on completed work not on projections. It could therefore be ascertained and checked with reasonable accuracy. The figures for returns proposed by the R and D people would be checked, and if necessary corrected, by those who had used the work – production people in the case of process improvements, and marketing people in the case of product innovations.
>
> Studies would be restricted to projects in which R and D had been pivotal.

The estimates of benefit made in this way would be on the conservative side. General rules for calculating them were laid down as follows:

> In valuing savings on process costs, minimum estimates based on savings on raw materials, energy, etc, are used, rather than higher values based on sales proceeds. The actual saving in each of the first five years is used. The five-year period is felt to be appropriate for the type of business but is admitted to be to some extent arbitrary.
>
> For one-off benefits, the actual value of the saving is used. An example of such a saving would be the individual application of a recently-developed technique for modifying the combustion conditions in a furnace in order to control NOx emissions into the atmosphere, thus saving the capital cost of adding extraction equipment.
>
> For new products, the value is assessed at 'prime margin' (price less directly variable costs, or value added less direct labour) for each of the first five years, with successively lower contributions down to zero over the next five years. Again, there is some admitted degree of arbitrariness in choosing the period over which to credit the benefit.
>
> For replacement products (ie where an older product had been withdrawn when a new one was introduced) an allowance is made, again on a prime margin basis, for the business displaced.

In addition to these benefits arising internally, Pilkington's R and D has also given rise to licensing income which is credited to it as part of the returns.

STRATEGIC MANAGEMENT ACCOUNTING

We have seen that the application of management accounting to intangible assets or revenue investment takes us some way towards realising the aim of demonstrating – as opposed to asserting – that such investment really does contribute to the long-term value of the firm, despite what the financial accounts may say. Attempts are currently being made to develop more systematic and comprehensive approaches to bridging this gap between what modern investors (see Chapter 1) require and what accounts produced for stewardship purposes can supply.

The financial aim of a business was defined in Chapter 1 as the maximisation of the net present value of future cash flows discounted at the company's cost of capital. The financial accounts are of no help in this endeavour since they reflect the past rather than the future. As we saw, concern about the immediate effect on the financial accounts may conflict with the aim of maximising the long-term value of the company, particularly when revenue investment, which will reduce the immediate profit in the interest of enhancing future returns, is involved.

The extension of conventional costing and investment appraisal techniques to the intangible realms of training, marketing, and innovation is perhaps the first step in a more comprehensive approach to management accounting, known as strategic management accounting (SMA). SMA attempts to use the concepts of management accounting to produce a comprehensive framework for guiding the strategic financial management (SFM) of the enterprise to produce the results which will eventually be shown in the financial accounts. It combines the comprehensive approach of financial accounting with the forward-looking emphasis of management accounting.

One of the aspects considered in SMA is the company's position relative to its competitors, particularly in terms of the competitiveness of its products and their production costs. In the present state of development SMA takes various forms and not a great deal of information is available because of the sensitive nature of the information involved (Bromwich and Bhimani 1994).

Strategic financial management attempts to deal with the problem of bridging the gap between financial and management accounts by evaluating NPVs, which are additive, for parts of a business or aspects of its management, allowing for the interactions between them, in a way which enables the figures to be reconciled with the financial accounts. The strategic approach includes the unrealised gains or losses which do not appear in the financial accounts but which contribute to the value of the business. It is based on cash flow which is the one aspect of company financial performance which can be unambiguously measured.

Questions for discussion

1 Why does the linear model of innovation, despite its shortcomings, predict a cumulative cash flow curve of the form of Figure 4.2?
2 How does the NPV of an innovation project change as the project proceeds?

Chapter 5

Financing an innovative new business

INNOVATION AND SMALL COMPANIES

Although innovation is a vital and integral part of the operations of any company, it is particularly associated in the public mind with new and small businesses. Under the largely unconscious influence of the linear model of innovation, much attention is given to encouraging and assisting the setting-up of new small companies to exploit academic inventions or discoveries and various studies have been made which claim to show that small companies are the source of a disproportionate amount of innovation.

Small companies certainly have the advantage of being able to respond quickly to events and to adapt readily to new ideas and opportunities, but they also have special problems, especially with raising the money with which to get started. Most new small companies are started with insufficient capital making them very vulnerable to any delays in getting a product on to the market and in receiving payment by customers. These problems are often magnified by reluctance of suppliers to extend much credit to a new and untried entity.

These problems are common to all new small companies, but they are magnified if the company is based on technological innovation since this adds an extra dimension of uncertainty to an already uncertain situation. However promising the innovation on which the business is to be based may be, much of its success will depend on business and managerial competence. A completely new product, if it has any value to potential users, will probably sell at first to a limited market; but if the market is promising, competitors will soon be drawn in and success will then depend on marketing and production skills as much as on technical ones. Luck also plays a part, since the design possibilities are many, and the winners emerge through an essentially evolutionary process (Dosi *et al.* 1988).

The pocket calculator is a good example, where the UK pioneer, Sinclair, was soon displaced by Japanese manufacturers whose products were more competitive on the combination of price, reliability, and easy availability. The story was later repeated with personal computers, where the Sinclair emphasis on small size and low price, attained at the expense of providing only minimum facilities, proved to be less attractive than a marketing strategy based on providing more expensive machines which had integral display and storage facilities, rather than relying on auxiliary use of domestic TV sets and tape recorders.

For a new company in an existing line of business, such as a new restaurant, the problems are mainly those of managerial competence and marketing skill. For a new company offering a new product in an existing market, as was arguably the case with the electronic calculator in competition with the slide rule, there are additional problems of acceptability of the product. If the product is a new one in a new market, then the problems are even greater. The consolation in such cases is that the competition will probably be less; the real difficulty is in defining or finding a market. Since this will probably be small, at least at first, there may be no special disadvantage in the company supplying it being small – indeed, there may be no scope for a larger one, at least initially, but the risks are high.

All these factors will weigh with potential suppliers of finance who will be influenced by the considerations discussed in Chapter 1. The high risk will make them look for a substantial return on their (or, more usually and crucially, their beneficiaries') money. Banks will lend at high interest rates, if at all; equity investors will want a large part of the business; and venture capital investment agencies will be looking for rapid growth, after which they will be able to sell out.

In extreme cases, where the risk is very high but some national advantage may be convincingly argued to exist in taking it, the government may assume the role of financial backer. Nuclear power and supersonic civil aircraft are the two obvious cases. The issues here are, however, more relevant to Chapters 6 and 7 below, and discussion of them is deferred until then. More modest support, by way of grants, is of relevance to the new business and is discussed in the final section of this chapter.

Finance and the small company

Although undoubtedly one of the principal problems facing a new business, that of raising money is not essentially different for one based on innovation than for any other start-up, although it may be rather more severe. We shall therefore consider only some of the special aspects relating to innovation. Any reader seeking guidance on setting up a new business will find many other sources of information on the general topic.

Both such readers and those with a more detached interest in the problems of setting up a new innovation-based company, and particularly those of raising finance for it, will find them very well described by the novelist Nevil Shute in his autobiography (Shute 1954). Shute was an aeronautical engineer by profession and, after an initial period after graduation working as an employee first on aeroplanes and then on airships, was one of the founders of Airspeed, one of the comparatively large number of aeroplane manufacturers set up in the UK between the two world wars. Accounts of the early struggles of small businesses set up more recently can be found in Bruce (1986).

It is possible for a single individual to set up a new business as a sole trader, or with one or more others as a partnership, but the advantages of establishing a company are considerable if the trading is envisaged to be on anything but a very small scale and it is intended to employ others in the business. These advantages, explained in Chapter 1, are essentially continuity and limited liability. In return, there are some formalities to be completed and some obligations in terms of reporting and accountability which do not apply to sole traders or partnerships. It is assumed in what follows that the founder or founders of the business have set up a company, probably made some sort of start, and are looking for extra funds to develop the business.

Any sources of finance, other than the founders of the company and perhaps their immediate relatives, will wish to assure themselves that the company seeking funds has adequate management in place, together with the record keeping and other systems needed to run the business. The basic document required for raising money is the business plan, which sets out the projected expenditure and income for a period of a few years, explaining why the money is needed and what it will be used for, together with evidence supporting the figures. This evidence will deal with markets, the people in the business, and the technology.

The underlying pattern of the business plan is likely to reflect the cash flow curve for an innovation shown in Figure 4.2. Translated into the year-by-year net cash flow figures (Table 4.1), the pattern is as shown in Figure 5.1. In this form, the eventual decline to zero of the income from the innovation shows up more starkly than in the cumulative curve of Figure 4.2, but it would be a brave and unjustifiable act to prepare an initial business plan which looked so far into the future.

The immediate problem for the fledgling company will be to fund the initial phase until break-even is reached, but the need to provide for a second generation of product to maintain the business once the initial product begins to lose its market appeal is something which should be borne in mind from the start. How soon the development of the second generation should start, and whether it should be a new product or an upgrade of the first, depend very much on the type of business and the timescales involved.

Ideally, a business should have enough initial capital to make a start on a successor product very soon after work on the first is under way, but this ideal is seldom achieved in practice since providers of capital will want to limit their exposure in the very early stages and in any case the amount of effort available is likely to be limited. Concentration is probably the best course of action in the beginning, but not to the exclusion of consideration

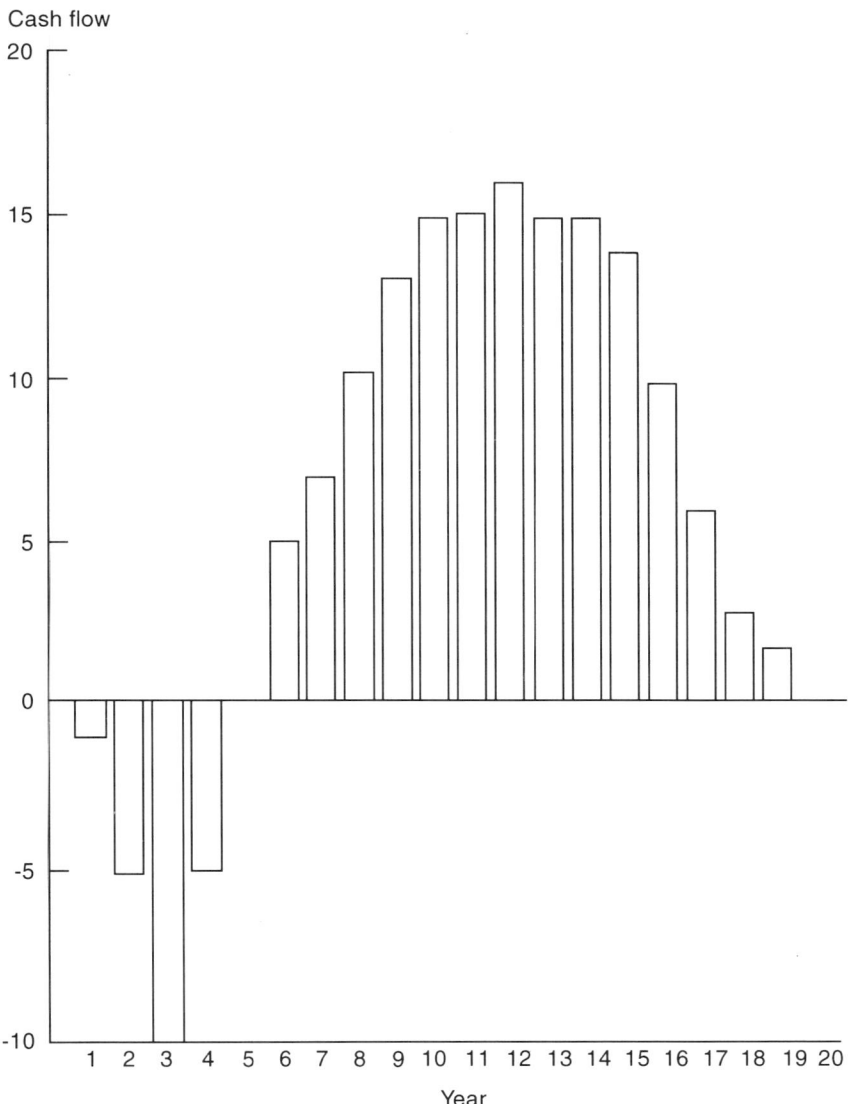

Figure 5.1 Annual cash flows for an innovation
Source: DWB

of what is to come next if the business has any intention of surviving in the long run.

To survive the difficult early years, and to provide a sound foundation for the longer-term future, a new business needs to have a limited number of committed and patient shareholders.

Finance and control

Founders of small firms, who have to be strong-minded and independent in character if they are to survive, are often reluctant to raise capital by selling shares to those outside their immediate circle fearing that loss of control would follow such a sale. This fear is largely, although not totally, unjustified: once there are outside shareholders, a certain amount of freedom is undoubtedly lost. On the other hand, it is rare for a company to be able to grow to more than a modest size without recourse to outside funds on a scale which can only be obtained through sale of equity. In any case, the independence of the proprietors of a small company which is heavily dependent on bank loans to stay in business is more apparent than real.

In some cases, founders of small firms have no ambition to grow their companies beyond the size at which they can satisfy the founders' ambitions for material and intellectual satisfaction. Such 'lifestyle' businesses are particularly likely to be found in niche technological markets. A company with greater ambition may find that if the shareholders are carefully chosen they may well be able to add more to the business than money alone, through advice, contacts, and guidance.

Companies which do not want to part with equity have two sources of outside funds, loans and their relatives hire purchase and leasing, and grants. We consider the first group below, leaving the rather special subject of grants until later in the chapter.

SOURCES OF LOAN AND SIMILAR FINANCE

Banks

Banks are the first port of call for those seeking money to set up a new business. The basic form of bank lending is the overdraft, which is repayable on demand and is secured against the company's assets. An overdraft has the advantage that interest is paid only on the amount outstanding at any time, on a day-by-day basis. It is really designed only to provide working capital or temporary funding for special purposes, although in practice many small businesses rely on bank overdrafts to a much greater degree than this. A survey in 1994 found that banks accounted for 60 per cent of the external finance of small firms in the UK (Bank of England 1995). Comparable figures for Germany are believed to be around 14 per cent (Bank of England 1994).

The lending bank will put a conservative value on the company's assets since they will be of interest to it only in the case of company failure when it will have to recoup its loan by selling the assets, necessarily under conditions where only poor prices are to be expected. The bank will also require that the directors of the company have a substantial stake in it themselves, and may well also require personal guarantees from the directors in addition to any security on the company's assets. The limit on a director's liability as a shareholder may thus be considerably increased in the role of director. (For completeness, it may be mentioned here that a director may also incur extra liabilities under the Insolvency Act if the business has been continued after it became clear that it had no reasonable prospect of paying its debts and it subsequently goes into liquidation. Limited liability applies to shareholders, not to directors.)

Banks are increasingly assessing loan applications in the light of the ability of the business to pay the interest out of cash flow, rather than on a purely security basis, and they are continually widening the scope of the services which they offer (Bank of England 1994). In particular, they will make fixed-term loans as well as overdrafts. Some banks have special departments devoted to serving innovative companies, with staff who have special interest in and knowledge of the problems of small companies seeking to exploit new products.

Fixed-term loans are less flexible than overdrafts: as well as the term, the amount is also fixed, and interest, which may or may not be at a fixed rate, has to be paid on the full amount of the loan for the full term. On the other hand, they cannot be called in at short notice, which is a great advantage for a business which will undoubtedly in most cases be permanently short of cash.

The government operates a Small Firm Loan Guarantee Scheme as an extra safeguard for banks lending to viable small businesses without a track record or adequate security to offer. The volume of lending sanctioned under this scheme tripled from £52 million in 1992/93 to £155 million in 1993/94 (Bank of England 1995).

Hire purchase and leasing

Companies seeking finance for specific pieces of capital equipment, particularly computer-controlled machine tools or other specialised equipment, may well find hire purchase or leasing a satisfactory means of financing their purchase. The firms which provide such services have specialist knowledge of the performance and applications of these forms of machinery, and are better able to cope with assessing the viability of a proposition than is the average bank.

SOURCES OF EQUITY FINANCE

For those who are willing to cede some formal control to outsiders in return for access to extra funding, sale of some part of the firm's equity shares provides the route. There are two main, interrelated, considerations: the policy decision of what proportion of the shares to sell, and the practical question of where to find the investors willing to buy those shares.

In strict company law terms, more-or-less total control of a company can be retained with a 75 per cent shareholding, since that is the percentage requirement for passing resolutions of major significance to the company's structure or objectives. The London Stock Exchange, however, requires that normally at least 25 per cent of the shares in a listed company must be held by the public. A 51 per cent shareholding is adequate for control over most activities, and in practice a holding of as low as 30 per cent is often regarded as sufficient. Much depends on the degree to which the holdings are spread amongst different owners with different views: none of the three holders of 30 per cent of the shares in a particular small company could be said to have control on the basis of shareholding alone, but the holder of 30 per cent in a larger company in which the next biggest shareholder was below the 3 per cent threshold – at which the holding would have to be made known to the company under company law – equally clearly would be. The 30 per cent figure is that at which the City Code on Takeovers and Mergers requires a holder to make an offer for the remaining 70 per cent.

Business angels

The term 'business angels' has come into use to describe wealthy individuals who are willing to invest in new companies, usually joining the board of directors and taking an active part in the management of the company. Shute (1954) found in the 1930s that such persons had often made money through land sales, but they are more likely to be found in the 1990s amongst those fairly senior in years who have retired from a major company after reaching a high level. The management experience and contacts which they bring to the new company, plus the implicit assurance of its respectability, are at least as important as the financial backing which they provide.

Business angels will be involved in providing sums up to around £100,000, rather more than a bank overdraft for a new company, but not enough for the foundation of a company with serious innovative ambitions. Such companies must look to venture capitalists in the first place and subsequently to the stock market.

The major difference between business angels and other providers of funds, apart from their individual knowledge and commitment, is that they are investing their own money, not that of other people to whom they owe fiduciary duties. They are therefore able to take greater risks, knowing that they will bear any losses themselves and cannot be held to account by anyone else.

Entrepreneurs seeking money are, understandably, much more concerned with their own needs than with those of the providers of funds, and hence sometimes less than sympathetic to the caution with which institutional investors must display in investing someone else's money. Nonetheless, as it is such funds who control the bulk of money available for investment, it is to them that the growing business will almost certainly eventually turn.

Venture capital funds

Venture capital funds invest money on behalf of others in speculative ventures of one kind or another. They are often specialist offshoots or in some other way connected with major financial institutions such as pension funds, insurance companies or merchant banks, on whose behalf they invest in high risk ventures in the hope of a high return. The normal pattern is for the venture capital investor to sell out after a comparatively few years, say five, taking the capital gain to invest in the next venture.

In order for the venture capital investor to sell out easily, the company must be a public company (plc) which must, by law, have a minimum allotted share capital of £50,000. Most venture capital funds prefer to invest in management buy-outs or other new enterprises in which the risks are more commercial than technical or innovative, but there are some funds which specialise in innovative companies. The members of the Venture Capital Association invested about 7 per cent of their funds in start-up and early stage companies in 1992 and about 6 per cent (amounting to £69 million) in 1993 (Bank of England 1995).

For over sixty years, there has been concern that companies are at a disadvantage when seeking funds on a scale which is larger than can be provided by individuals or bank overdrafts, but is small by the standards of the major investors such as pension funds which have millions of pounds per day of new money to invest. The cost to the prospective professional investor of examining a proposition – a process which is known as 'due diligence' – is much the same, whatever the size of the investment within broad limits, and these costs must be considered in relation to the size of the investment. It is clearly not justified to spend a substantial fraction of the proposed investment on deciding whether or not – especially as the answer is much more often not – to make the investment.

This financing gap, which became known as the 'Macmillan gap' after the chairman of a committee which reported in 1931, is in current terms around £150,000 to £3,000,000. An approach to closing it is reported in HMG (1995), which says that the government is to pilot a specialist pre-finance evaluation service to help smaller businesses overcome the difficulties which they have in raising medium and long-term finance for innovative and 'high-technology' projects. (The term 'high technology' or sometimes 'high-tech', is often used rhetorically, and is more often used than defined. Definitions used by OECD, quoted on page 35 of HMG (1995), are that 'high technology' manufactures have R and D expenditure of greater than 3.5 per cent of sales, while 'leading edge' manufactures have R and D expenditure of over 8.5 per cent of sales. Curiously, HMG (1995) itself uses the much more satisfactory term 'R and D-intensive' at this point. In the context of the proposed new service, 'high technology' projects are probably those in which the technology is new and is a major factor affecting the project risk.)

An earlier approach to filling the 'Macmillan gap' was the setting up in the

1940s, by the Bank of England and the major clearing banks, of the Industrial and Commercial Finance Corporation (ICFC), which had a specialist arm known as Technical Development Capital (TDC) for investment in technology-based new companies. ICFC, with TDC, subsequently evolved into the company known as 3i (Investors in Industry).

Venture capital trusts and similar schemes

Venture capital trusts (VCT) are a new form of investing institution which combine some of the features of business angels and of venture capital funds. VCTs were first proposed in the autumn budget of 1993 with further details following in the autumn budget of 1994. Rules providing for their admission to listing on the London Stock Exchange were adopted in August 1995. They are especially aimed at the needs of small companies, who will be allowed to raise up to £1 million per year from a VCT. As the name suggests, VCTs will spread their risk by investing in a number of companies. The VCTs will get their money from individual investors who will buy shares in the VCT. For investments up to £100,000 per year, the investors will have exemption from income tax on dividends and exemption from capital gains tax on disposals. Investors who buy new shares and keep them for five years will also be entitled to income tax relief at 20 per cent and to defer tax on capital gains where the gains are reinvested in a VCT.

The VCT rules are designed to minimise the risk to individuals of investing in smaller companies and to encourage them to keep their funds invested in the smaller firms sector. Similar rules apply to other schemes, such as the Enterprise Investment Scheme (EIS) introduced in the November 1993 budget to replace an earlier Business Expansion Scheme, and modified in the November 1994 budget.

The EIS is far from unusual in undergoing rapid modification which is a common feature of government schemes, all descriptions of which tend to date very quickly. The details of the individual schemes are important to their users at the point of use but, from the more general standpoint of this book, the details are less important than the fact that official encouragement is being given to investment in small firms, and that their special difficulties in introducing innovative products have been recognised as worthy of some degree of relief in addition to the support provided by grant schemes discussed below.

The stock market

A stock exchange fulfils two functions. First, it provides a means of raising new capital for companies by the sale of shares and, second, a market in which those shares can be traded. There is a considerably greater amount of activity in the second of these functions than in the first. As was pointed out in Chapter 1, one of the attractions of an investment of shares in plcs quoted on the stock market is that they can easily be sold or bought – admittedly at a price which depends on the balance of supply and demand and may vary considerably from day to day – thus providing the shareholder with liquidity while leaving the company in possession of the original funds. From the point of view of the growing company, it is the capital-raising activities of the stock market which are most important, while for a mature company it is the effects of the workings of the secondary market. These effects, particularly on innovative activities, are discussed in Chapter 6 below.

To protect investors, the law relating to offering shares for sale to the public lays down very strict rules. A company seeking to become listed on the London Stock Exchange has to provide listing particulars (known as a prospectus) which give very considerable detail about all aspects of the company, its past, its present, and its future prospects. The Financial Services Act lays down that the listing particulars should contain:

> ... all such information as investors and their professional advisers would reasonably require, and reasonably expect to find there, for the purpose of making an informed assessment of:
> the assets and liabilities, financial position, profits and losses, and prospects of the issuer of the securities; and
> the rights attaching to those securities.

Under the Act, the detailed implementation of the rules is delegated to the London Stock Exchange (under its official title of the International Stock Exchange of the United Kingdom and the Republic of Ireland Ltd), whose Listing Rules are contained in the 'Yellow Book', an A4 loose-leaf compilation some three centimetres thick. The Exchange normally requires that the applicant for listing has a sponsor, usually a broker or investment bank.

Navigating through this minefield requires extensive, and expensive, professional advice and assistance. (Gower (1992) refers in a footnote to issuers being 'appalled' by the proposed charges of their first choice of adviser, only to find that rivals were no cheaper.) The effort required from management is also considerable, and managements have been known to find the whole process so distracting that the business was neglected to the point of danger while attention was devoted to the listing exercise (Bruce 1956, chapter 2).

Despite the difficulty and expense of obtaining and then maintaining a listing, and apart from the fact that venture capital investors may insist on it, provided that the company can demonstrate the three years of successful trading which are required as a condition of listing, there are very considerable advantages to be gained. These are:

> Access to substantially larger amounts of new capital than can be obtained otherwise. (The London Stock Exchange requires that the expected market value of the securities to be listed should be at least £700,000.)
> If the company proves to be attractive to investors, as technology-based

companies often do, the resulting high P/E ratio means that the capital raised will be comparatively cheap. An alternative way of looking at this phenomenon is to note that it demonstrates that the company has built up a considerable amount of intellectual capital.

Improved standing in the eyes of customers, suppliers, and others. Although intangible, this benefit may have significant effects on the business.

The original founders of the business, as well as the early outside investors, can release some of their wealth from it, either as a form of insurance or to provide for retirement or some other personal need.

From the investor's point of view, admission of a security to the Official List means that there will always be someone, known as a market-maker, who will buy, or sell, shares at the current price at any time. The continuing obligations on a listed company in terms of provision of information at regular intervals and about any significant development also protect the interests of the investor.

Investors in listed securities are usually looking for a reasonably safe and predictable return which will increase at least in line with inflation, and preferably in advance of it. Despite this preference, and against advice, the directors of Oxford Instruments felt that the requirements of a prospectus obliged them to include in the one which they issued in October 1983 the policy declaration: 'To meet our objectives in the longer term, substantial investment in product development is intended and necessary and may sometimes take priority over the demands of short-term profitability.'

Oxford Instruments remains, in relative terms, one of the heaviest spenders on R and D in UK industry and the policy which it espoused in 1983 is perhaps now regarded as being more understandable and acceptable than it was at the time.

Scientific research-based companies

In December 1993, the London Stock Exchange introduced special rules for the listing of scientific research-based companies. These rules were updated in January 1995 and are intended particularly for companies involved in the development of chemical and biological products or processes, including pharmaceutical companies, and those involved in diagnostics, agriculture, and food.

The rules were designed primarily to meet the special case of pharmaceutical and related companies in which the time scales of development of new products and of the generation of revenue from them are very long, and the investment needed before that return is gained is very substantial. With the agreement of the Exchange, other similar innovative science-based companies, such as another embryo Oxford Instruments, may also take advantage of the rules.

A company seeking to take advantage of the rules for scientific research-based companies must be able to demonstrate that it has attracted funds from 'sophisticated' investors, and that it intends to raise at least £10 million on top of an original capital of at least £20 million to bring identified products to the point at which they can raise significant revenue.

The main difference from the normal listing rules is that the requirement to demonstrate three years of revenue-earning activity is replaced by a requirement to show at least a three-year record of successful R and D, resulting in patent applications, tests or trials establishing the viability of projected products; and in various other ways to demonstrate the prospect of developing into a viable business.

The special listing rules apply a test of achievement of commercial milestones such as having at least two drugs in clinical trials, the conclusion of a development agreement involving the receipt of at least £5 million and the retention of a significant financial interest in any intellectual property arising from the development, or the expenditure of at least £20 million on R and D over at least three years which has resulted in the generation of intellectual property of significant value which the company intends to develop as its main business activity.

There are also rules intended to give prospective investors reasonable reassurance that the directors and the major investors in the company are likely to remain with it both physically and financially, and that the claims for the prospective products are supported by independent assessors.

Junior markets: USM and AIM

From time to time, markets have been set up to deal in the shares of smaller companies with fewer rules and requirements for the companies and correspondingly greater risks for investors than those existing in the major market. Such markets provide less for both parties than full listing, but have been found appropriate by some members of both the investing and fund-seeking communities.

The London Stock Exchange set up the Unlisted Securities Market (USM) in 1980 to cater for this group of clients whose numbers had increased and who had been satisfying their needs in a variety of ways. The rules of the USM were based on those of the Official List, but were relaxed in some particulars. The USM was popular for some years, and was often the first stage on the way to full listing. Smaller companies wishing to 'go public' often went first for the USM and then, if they grew successfully, transferred to the main market. Nearly 900 companies came to the market through the USM, and they raised nearly £6 billion, but interest in it declined after the 1987 stock market crash (Chapter 1) and the subsequent recession.

The USM was also adversely affected by increasing requirements for investor protection, partly stemming from EC harmonisation measures, and the London Stock Exchange took the decision to close it at the end of 1996. In anticipation of this event, admission to the USM was closed to new entrants on 31 December 1994.

The original intention of the London Stock Exchange was not to replace the USM, but it eventually decided to introduce the Alternative Investment Market (AIM), which opened in June 1995, initially with ten companies who were soon joined by others. Membership of the AIM is likely to be further increased when the USM closes, leaving its remaining members with the choice of transfer to the Official List, to the AIM, or losing a quotation altogether.

The AIM caters for companies without a trading record and is expected to attract smaller biotechnology and other technology-based companies; companies which have a mainly regional presence, and family companies seeking a quotation to ease liquidity or enhance prestige.

SOFT MONEY FOR INNOVATION

'Soft money' is a term, probably of US origin, to describe grants from public sources and similar funds which are obtained by essentially political processes rather than on a commercial basis. The essence is that the provider of the soft money is seeking an indirect return, such as a stronger economy, higher employment, or the improved exploitation of publicly-funded research, rather than the prospective interest, dividend, or capital gains which motivate the provider of commercial funds. At the working level, the provider of funds will be an official who has rules to apply, targets to achieve, and as little in the way of direct personal experience of the world of the recipient as the recipient has of that of the official.

Obtaining soft money is therefore a different business from obtaining commercial funds. The arguments which have to be deployed are different but, while they tend to remain much the same from time to time in the commercial world, they may vary quite quickly in the political field. Political priorities change, not merely from party to party if there is a change of government, but from minister to minister and even from official to official. Adaptability and a keen awareness of current trends are therefore needed on the part of those seeking funds. The preferred topics, methods, and types of company can vary from time to time. Current themes are the encouragement of small firms and of academic-industrial collaboration.

General principles

Despite the tendency of government support schemes to change with a rapidity which is bewildering to those not in day-to-day touch with them, some principles remain fairly constant. The most important of these is that public support will normally be limited, usually to 50 per cent of the eligible costs. This is an EC rule, aimed at controlling competitive conditions between the member states by limiting competitive subsidy. It is in any case soundly based in the more general managerial principle that financial involvement concentrates the mind. It also helps to counteract the perfectly natural tendency of technical innovators to aim for the most adventurous and technically-complete product, regardless of cost.

Shute (1954) describes this process in his first-hand account of the interesting and unrepeated experiment in which two rival airships were constructed to the same specification, one by private industry under a fixed price contract (R100), and the other directly by government with no such limitations (R101). The commercial pressures on the private constructor produced a notably better result (R100 successfully made a voyage to North America and back) than the combination of the absence of such pressures and their replacement by political pressures on the rival craft, whose unfortunate fate has gone down in the history of air travel.

The lesson was, however, either not learned or was forgotten in the aftermath of war until it was relearned with the supersonic aircraft Concorde, perhaps the outstanding example of the tendency of technological projects to get out of hand unless anchored firmly to market realities. The argument for Concorde was that it was the next technical step to take after the introduction of the jet engine into civil aviation, and that all previous increases in aircraft speed had been commercially worthwhile. This argument was deployed from a government laboratory and, after a good deal of political manoeuvring, the Anglo-French agreement, signed in 1962, provided that the development should be entirely financed by the two governments, albeit carried out by companies. The result was a technological triumph but a commercial disaster.

A second principle applied by the UK government is that of additionality. The government will not support projects which would have gone ahead without that support. It is difficult to see how any limit could be put on subsidy without the additionality principle, but its application is clearly a matter of judgement and, in particular cases, sometimes causes difficulty and even resentment. Potential recipients of grant point out that the principle tends to discriminate in favour of marginal projects, to which government replies that the purpose of a grant scheme is not to subsidise work which would be done without subsidy. It is intended to enlarge the scope of work done by reducing the risk, reducing the time taken, or promoting collaboration which would not have taken place without support. The difference is one of viewpoint, not substance: additionality is indeed marginality, but it is not necessarily the worse for it.

One consequence of the additionality principle is that no contribution by way of grant can be made retrospectively. The details of the project to be supported must be agreed in advance, and no expenditure incurred before the official start date can be included in the eligible costs.

The definition of eligible costs can vary from scheme to scheme, particularly as regards the figure allowed for overheads. Some schemes have a fixed allowance for overheads, usually expressed as a percentage of labour costs, while some permit negotiation on this point.

In any collaboration, even one as simple in principle as a grant, it is essential to adhere to the principle of agreeing in advance what the terms of the collaboration are. The results to be achieved or the targets to be aimed at, the respective responsibilities of the parties, and the ownership of any fruits of the collaboration which may appear, are the absolute minimum on which agreement should be reached. Not only does adherence to this principle reduce the scope for disagreements when problems arise, particularly about the ownership of IPR, but the process of reaching the agreement is beneficial to the project, as it ensures that most aspects are thoroughly considered in advance.

Collaborative projects

The last point applies with particular force when the project is of the more general collaborative kind in which more than one partner will actually be carrying out work. Even between two companies in the same country there may well be substantial differences of approach and general culture which can hinder communication and hence inhibit successful collaboration, and the problems become more pronounced when the participants are from different sectors of the economy or from different countries.

It is especially these two kinds of collaboration, particularly academic-industrial collaboration through schemes such as LINK, and the increasingly significant EC schemes under the Framework Programme, to which the majority of generally available funds are nowadays devoted. Participation in these schemes involves fulfilling the requirements in terms of the number and distribution of collaborators which the individual scheme requires. It is for each potential participant to judge whether any slight artificiality which results, and the extra effort and expense which are undoubtedly involved in forming and carrying out a collaborative project, outweigh the financial contribution from public funds and the equally undoubted benefits which access to external expertise and contacts can bring.

One intangible and often unconsidered benefit of involvement with a publicly-funded scheme is that successful projects may obtain a good deal of free publicity through publications about and presentations on the particular scheme, the promoters of which will both be keen to demonstrate its success and have opportunities to do so.

The benefits of collaboration can, of course, be obtained by direct collaboration without the involvement of a support scheme. A small firm with a new product may well find that the full exploitation of that product is beyond not only its current resources but of any resources to which it might reasonably aspire to have access. Perhaps the most conspicuous example is the pharmaceutical industry, in which the testing and certification procedures necessary to get some types of new therapeutic entity on to the market are so expensive and time-consuming that only large companies with established businesses can hope to finance them. An example is discussed in Chapter 6 below.

It is in the interests of both the large pharmaceutical companies – always aware that their success is largely dependent on a very few products, that a constant stream of new ones is required to sustain the business, and that they cannot pursue all the possibilities which exist – and the small biotechnology companies – possessing specialist and possibly world-leading expertise in particular niches but lacking the funds to pursue any major exploitation opportunity – to collaborate, as indeed they do. The conditions of the Listing Agreement of the London Stock Exchange, described above, specifically recognise such collaborative agreements as helping to establish the suitability of immature biotechnology companies for admission to listing.

The pharmaceutical industry is by no means alone in offering opportunities for collaboration which may be of assistance to a small and growing company. The current tendency is for major companies to restrict the scope of their direct in-house activities to what they regard as their core functions and to contract out many necessary, but less specific, functions. Because of the need to develop new products quickly and to a very high standard of quality in order to compete in an increasingly global market, the contracting-out is increasingly of a collaborative nature, rather than by way of straightforward orders or contracts to supply, thus offering possibilities for assisting the development of the supplier. The automobile industry is one in which this development is well advanced, with formerly adversarial relations with suppliers being replaced by collaborative long-term partnerships.

Some members of the retail sector, notably Marks and Spencer, have long had a policy of working in close collaboration with their suppliers, whose capabilities have been much assisted by the process.

Specific small firms' support schemes

The UK government has for some years pursued a policy of not supporting 'near market' projects, particularly in R and D. This policy is essentially an application of the additionality principle, but it has sometimes been applied from an unrealistic view of what the market would in practice support. Small firms are, however, exempted from the full rigour of its application, and support schemes for near-market product development have been set up. The details of each scheme, and indeed the existence of the individual schemes themselves, can change rapidly, but the general possibilities for their design are relatively limited, and can be illustrated by brief descriptions of some current schemes.

SMART (Small Firms Merit Award for Research and Technology) is an annual competition which offers 180 awards at a rate of 75 per cent up to a maximum of £45,000 to carry out 6–18 month technical and commercial feasibility studies of innovative technology projects with commercial potential undertaken by businesses with fewer than fifty employees. To qualify, such firms also have to have either a turnover of not more than five million ECU or a balance sheet total of not more than two million ECU. SMART is designed to help such small firms over the first hurdle to the point at which they may be able to attract funds from commercial sources.

SPUR (Support for Products under Research) supports products under development, but was given a name whose euphony and implications are more in keeping with its objectives and aspirations than a more accurate one would have done. SPUR provides grants to small businesses with up to 250 employees to develop products or processes which promise significant technical advance for the industry concerned. It offers a flat rate grant of 30 per cent of the eligible costs, with a maximum grant of 200,000 ECU which is about £156,000. Winners of SMART awards who want to take their project further can apply for SPUR grants at the preferential rate of 50 per cent, but with the same maximum grant (in this case for both grants combined) as other SPUR applicants. There is also a SPUR-PLUS scheme, which supports projects of exceptional strategic promise. The grant rate is again 30 per cent, but the maximum grant is £450,000.

Approved funding for the three years 1995–98 is £69 million for all three schemes, of which only £3 million is allocated to SPUR-PLUS, with the rest being divided about one-third to SMART and two-thirds to SPUR.

The Teaching Company Scheme

The Teaching Company Scheme (TCS), while not limited to small companies, is a form of academic-industrial collaboration which is more generally suited to them than the other current such scheme, LINK. The TCS was started in 1975 to link industrial companies and higher education establishments in working partnerships aimed at making a substantial change in the techniques, products, or procedures used by the industrial partner. It specifically includes management aspects as well as technological ones. Each partnership is, confusingly, called a programme, and typically lasts for two or three years. The work is carried out by Teaching Company Associates who are employees working on short-term (usually two year) contracts under joint academic and industrial supervision. A typical programme will have more than one Associate over its life.

The initiative for a TCS programme comes from the company, and a central secretariat helps to find the academic partner if the company does not have one in mind. The government grant goes to the academic partner, and the industrial partner makes an annual contribution per Associate which depends on the company size, whether the partners have previously collaborated, and other factors, but is around £8,000–9,000 per year for a company with fewer than 250 employees and £14,000–15,500 for a larger company.

The budget of the TCS is around £15 million per year, and at the end of the 1993–94 financial year there were over 500 TCS programmes with a total value of around £70 million, of which £26 million was direct funding committed by industrial companies.

The LINK scheme

LINK (which is not an acronym) is also an academic-industrial collaboration scheme, aimed at 'pre-competitive' research, that is, research which is not near-market. ('Pre-competitive' is a term stemming from EC competition policy which frowns on collaboration between competitors except in special circumstances.) LINK was introduced in 1988 and is organised in Programmes, each with a theme and a number of appropriate financial supporters drawn from government departments (to support the industrial partners) and Research Councils (to support the academic partners). Within each Programme there are collaborative projects, each with a minimum of two and usually more partners, at least one of whom must be from the 'science base' (the publicly-funded research organisations, mainly in universities and government laboratories) and one from industry. Although there is special emphasis on involving small companies in LINK, it is suitable only for those which are research oriented.

LINK operates on a strict 50 per cent funding rule, which means that 50 per cent of the eligible costs for each project have to come from non-government sources, usually the industrial partners. Since the academic partners normally require that all their eligible costs should be met from the project budget, the amount left for the industrial partners depends on their share of the total eligible costs. On average, the result is to give the industrial partners a grant at around 30 per cent, but this figure can range from below zero (if the academic partners have more than half the eligible costs) to perhaps 40 per cent, above which figure doubts begin to emerge about how far the academic partners are really contributing.

Symbolically, the LINK funding formula can be expressed in the following equations where C denotes costs, G denotes grant, I denotes industry, A denotes academia, and T is the total:

$$C_T = C_A + C_I$$

$$G_T = C_T/2 = G_A + G_I$$

$$G_A = C_A$$

From which:

$$G_I/C_I = 1 - (1/2(C_I/C_T))$$

The grant to industry is thus zero at a cost share of 50 per cent, 25 per cent at a cost share of 66.7 per cent, 33.3 per cent at a cost share of 75 per cent and 37.5 per cent at a cost share of 80 per cent (Figure 5.2).

This feature of LINK illustrates the influence of cost accounting conven-

Grant rate

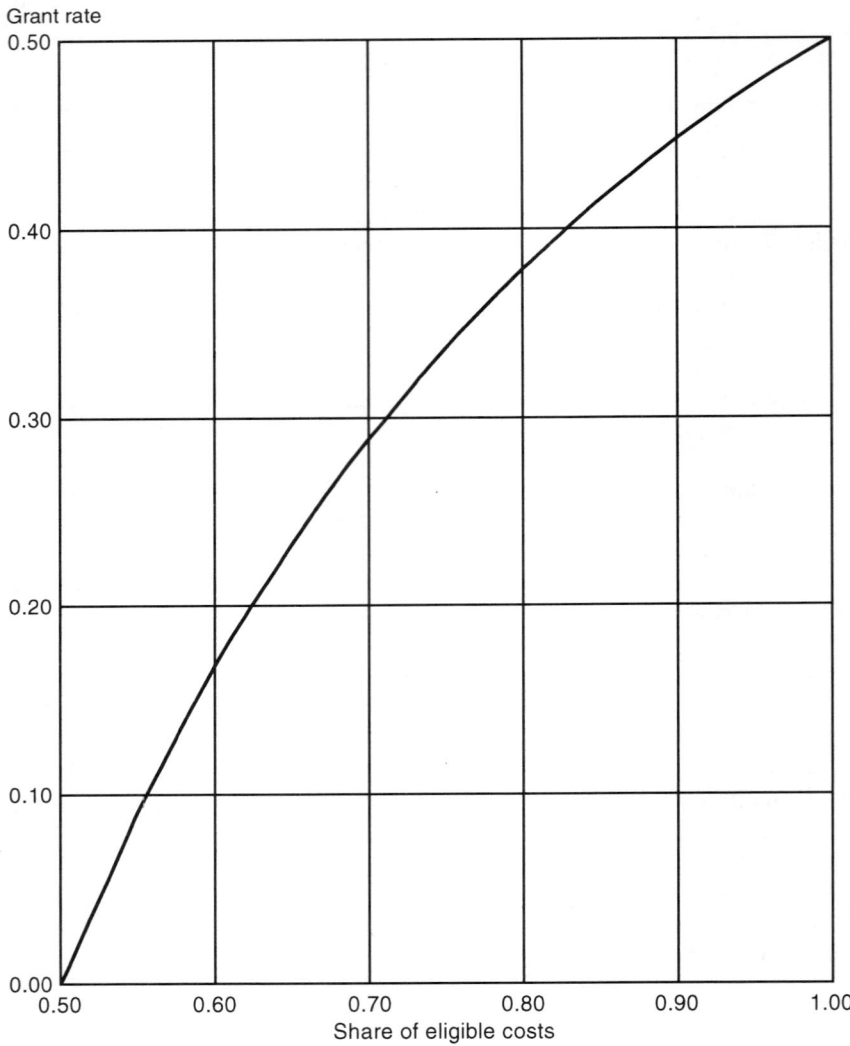

Figure 5.2 Industrial grant under LINK funding
Source: DWB

tions and regulation. Government accounting conventions in particular tend to appear somewhat strange to those used to industrial conventions, and academic conventions are different again. The eligible costs for the academic partners in LINK projects do not correspond to the real costs involved, with the result that the academics are not as cheap as they may claim to be, and the industrial partners in practice pay a smaller proportion of the real costs than appears from the figures.

This result arises because although the EC rule which restricts subsidy to 50 per cent of full costs applies equally to industrial and academic institutions, the EC recognises that no UK academic institution has a costing system which can yield such information, and therefore accepts that they can instead receive 100 per cent of marginal costs. Marginal costs cover staff specially recruited or employed, and direct expenses on equipment and materials. The costs of employing the permanent academic staff who may make a considerable contribution to a project, and the costs of premises and support services, are not included. The support which academic institutions receive for such costs, which is itself a controversial subject outside the scope of this book, are not eligible costs for LINK.

Individual LINK Programmes have grant budgets which average £7 million, making an average total programme cost of £14 million which is usually enough for around twenty projects. Programmes have a limited life for allocating money of around five years, and the dynamic equilibrium number of Programmes is around thirty.

Preferred topics for support

At any one time, the preferred topics for government support tend to be limited. Apart from defence, which is a specific government responsibility, and civil aerospace, which has a special place which is discussed further in the next chapter, emphasis tends to change from time to time. Nuclear power was a leading topic for a period, but is now regarded as mature enough to stand, or die, on its own feet. Governments around the developed world have tended recently to select information technology, biotechnology, and materials technology as the areas of special promise for the future which are in need of a helping hand from public funds.

The UK government's 1993 science, engineering, and technology White Paper (HMG 1993) launched the Technology Foresight programme as a more systematic means of attacking the problem of identifying promising areas for support. The first fruits of this programme emerged in 1995 through a series of fifteen reports from panels set up to study particular topics, and a report from the Steering Group for the Foresight process (OST 1995a).

OST (1995a) identifies a number of priorities for future work which will in due course influence the selection of topics for LINK programmes and the priorities of the work of the national science base. Apart from twenty-seven generic science and technology priorities, the report identifies eighteen generic infrastructural priorities, amongst which is included the development and encouragement of long-term finance for R and D and innovation, including the continuous review of fiscal measures. Discussion of this latter topic is taken up in the final chapter of this book.

Questions for discussion

1 Is the amount of money available from HMG in grants to support innovation significant in the context of other funds available? Is it likely to have effects other than on the recipients?
2 The 'Macmillan gap' has persisted for over sixty years. What are the implications?

Chapter 6

Finance and innovation in a mature business

THE DYNAMICS OF A BUSINESS

If all goes reasonably well with a new business it will become established to the point at which it becomes self-supporting. Depending on the opportunities and the ambition of those running it, it may continue to grow to a greater or lesser extent or may reach a size at which it remains over an extended period. Even in such cases, however, the equilibrium will be a dynamic one, with old products dropping out of production to be replaced by new ones. Growth or shrinkage are special cases of departure from this dynamic equilibrium state.

The current mental model of a company's operations, especially as represented by its accounts, makes very little acknowledgement of the need to replace old products with new ones. As we have seen, it is recognised, by the depreciation provision, that a company must replace, or at least make financial provision for replacing, its machinery and other fixed assets which wear out over time, but there is no such obligatory provision for continuously renewing its intellectual capital.

The problems of 'short-termism' to which this lacuna gives rise have been discussed above in Chapter 1. There is undoubtedly a conflict between maximising this year's profit and making appropriate provision for the future, and one of the major problems confronting the management of a mature business is how to resolve this conflict in the best long-term interests of the company. This is a major strategic issue and the approach to it will differ from company to company depending on its strategy. Some will favour an innovation-led strategy in which the company seeks to be the first in the field with new products, while others will be content to follow to some extent in their wake, seeking a competitive advantage through better marketing, superior production technology, or exploitation of an established reputation or other strategic advantage.

Whatever the strategic approach, however, the management needs to have some concept of the dynamics of the business which it is seeking to run, and to allocate its funds and attention accordingly. In sectors where innovation is expensive and time-consuming, such as in pharmaceuticals and aerospace, these concepts have been developed to a considerable degree of sophistication, aided by the fact that new products are quite distinct from each other and from their predecessors. In such industries, the scope for strategic choice is rather limited: a pharmaceutical company can choose to be research-based, basing its business on a succession of patent-protected new chemical entities; or it can rely on manufacturing and marketing out-of-patent medicines, licensing in new ones as opportunity offers; but intermediate strategies, involving a modest degree of research intensity, are not viable.

Glaxo Wellcome, SmithKline Beecham, and Zeneca are examples of the research-based strategy, while Fisons and Boots, which attempted to follow an intermediate course, have both abandoned it. The four components of Glaxo Wellcome and SmithKline Beecham were already large in the terms of the pharmaceutical industry, but merged to form even larger units, while Fisons and Boots were smaller and have effectively shrunk further. Fisons opted for the manufacturing and marketing strategy in 1994 and sold most of its R and D operation, while Boots chose to leave the field entirely and, in 1995, sold its pharmaceutical interests to a larger concern.

These examples illustrate the importance of size in such innovation-intensive industries: Fisons and Boots each had an annual turnover in pharmaceuticals of £450–500 million, and related annual R and D expenditure of £50–60 million while Zeneca, the smallest of the other three, had a turnover in pharmaceuticals of almost £2,000 million in 1994, and an R and D expenditure on pharmaceuticals of about £300 million, with a declared target of introducing one new therapy per year. The difference between the ways in which Boots and Fisons chose to deal with what both decided were sub-viable operations is accounted for by the difference between the contribution which their pharmaceutical interests made to their total businesses. Boots' pharmaceuticals constituted only about 10 per cent of its sales, whereas the figure for Fisons would have risen to almost 50 per cent after the planned disposal of another business.

In aerospace, the demands of innovation have resulted in an even higher degree of concentration on a world-wide scale, with only three makers of large passenger aircraft and three major engine-makers left.

At the other end of the scale, where innovation is comparatively cheap and rapid, as in the clothing and food industries, an innovation cycle can be accommodated within one year's trading, greatly simplifying the investment decision process. For sectors in which the characteristics of innovation are intermediate, with perhaps a several-year investment cycle and a long exploitation period to follow, with much of the available opportunity lying in gradual or incremental improvement, the decisions can be much less clear-cut than at the extremes. Much of mechanical engineering falls into this category, as do the various materials industries. In both the short- and intermediate-cycle sectors, the opportunities for strategic choice are comparatively wide, adding to the problems of management.

The business as a succession of innovations

In Figure 5.1 we looked at the annual cash flows from an innovation whose cumulative cash flow was of the established form shown in Figure 4.2, noting that it eventually came to an end. We also noted that if the business was to continue then it would need to introduce a new product, preferably before the old one had entirely ceased to produce revenue. Looking further into the future, this second product would itself, in due course, come to the end of its life, as would its successors. A business which is aiming at a continued existence therefore needs to introduce a succession of innovations, suitably spaced out in time, so as to maintain itself.

Figure 6.1 shows the annual cash flows for a business in which innovations with the cash flow pattern of Figure 5.1 succeed each other at intervals of three years. It can be seen that the period of negative cash flow is increased relative to the single-product case, but that after seven years the cash flow becomes positive and then builds up steadily until about year sixteen, after which it becomes stable. At this point and thereafter, provided the supply of innovations is maintained, the business will be stable.

Figure 6.1 also shows what happens when the flow of innovations is interrupted, in this case by starting no new projects after the ninth one, which begins in year twenty-five. The consequent reduction in innovation expenditure immediately shows up in increased cash flow, which continues to increase until year thirty, after which it drops, gradually at first and then precipitately as the cupboard becomes increasingly bare.

The underlying assumption in Figure 6.1, that all innovations have the same cash flow profile, is of course over-simplified. Nevertheless, the figure does illustrate the dynamic relation between innovation and finance, and distinguishes the various stages through which a business must pass on its way from the cradle to the grave. Thus:

> Stage 1 Early period of struggle, dependent on the original capital funding and any other early-stage funding. (Years 1–10)
> Stage 2 Period of steady expansion. (Years 10–16)
> Stage 3 Period of stability. (Years 16–27)
> Stage 4 Period of false growth obtained by starving investment in the future. (Years 28–30)
> Stage 5 Period of decline. (Years 31–44)

The first three stages are valid for all businesses while the fourth and fifth, although frequently observed, are not. The speed with which changes take place is different from business to business since it depends on the time scale of the innovation process for that business. In the case taken for Figures 5.1 and 6.1, this process involves an investment period of four to five years, followed by two years of small return, ten years of healthy return, and three to four years of decline. These figures are reasonably representative of the

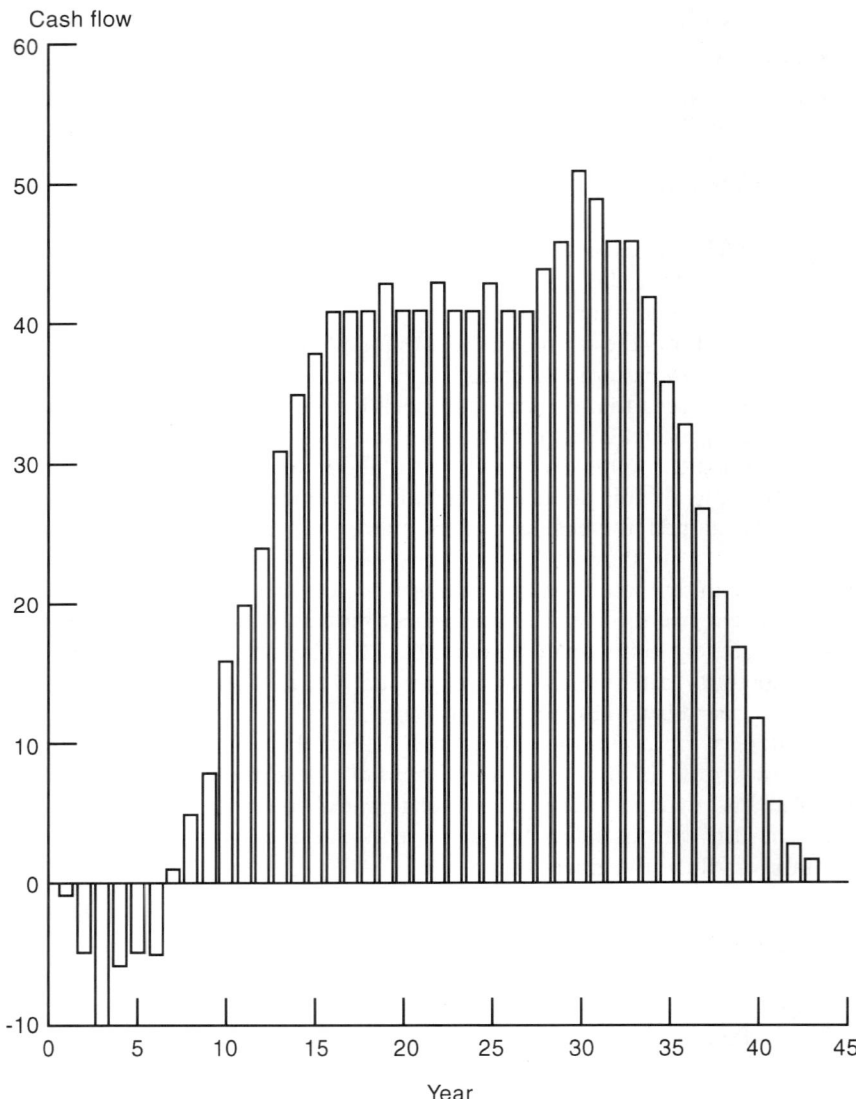

Figure 6.1 A business as a series of innovations
Source: DWB

mechanical engineering sector but would vary from case to case, being longer in pharmaceuticals and aerospace and shorter in consumer electronics for example.

The kinetics of a particular industry are very important, especially as it is much easier for an incoming management to cut expenditure on the future than to increase it in a fruitful way. This 'whizz kid' management approach, seen in Stage 4, can be seen to produce better than the sustainable results for a period rather longer (about seven years) than the investment time for a new project, and probably longer than the managers responsible would stay, leaving successors to pick up the pieces.

Figure 6.2 shows what happens if the second and subsequent innovation cycles are started at intervals of five years instead of three. It can be seen that both the early development of the business and its subsequent performance are distinctly erratic, compared to those of the business with more closely-spaced innovations. The importance of starting on the second product as soon as possible is clear, but the availability of funds will limit the extent to which this can be done in a practical case.

Complicating factors in a real business

Innovation expenditure – expenditure on the future – is, as the ASB (1993) recognises, discretionary. The employees and the suppliers have to be paid, as do the tax collector and the bank's interest, and the law requires that a provision for depreciation be made. Although this provision does not affect the cash available, it reduces the amount available for distribution to share-holders who, in a mature company, will expect a dividend. The directors of the company will be all too conscious of the fact that shareholders who are disappointed in this respect may either replace the offending directors by direct action or sell their shares in a takeover bid to another company whose first action will probably be to do that job for them.

In consequence of these pressures, directors are tempted, when times are hard, to cut expenditure on innovation in the hope that it can be resumed when trading conditions improve. The danger is that the lack of innovation will so damage the company that it will be unable to take advantage of an improvement in the general economic conditions, and the period of inevitable and ineluctable decline of Stage 5 sets in.

This is what happened to the UK motor cycle industry under the impact of Japanese competition in the 1960s and early 1970s (Boston Consulting Group 1975). The Japanese first introduced small machines which displaced their British competitors. Rather than improve their own machines in order to compete, the British manufacturers retreated into larger machines, in this way preserving their return on capital which was their strategic aim. Unfortunately, the Japanese manufacturers pursued them up the size scale until eventually there was virtually nothing left of the British industry.

The British-owned automobile industry followed much the same path a few years later, keeping models in production long after they had been overtaken by competitors. By the time commercial failure occurred most of the industry was in the hands of British Leyland, which was too large to be allowed to disappear, and was therefore taken into public ownership in 1975.

Cash flow

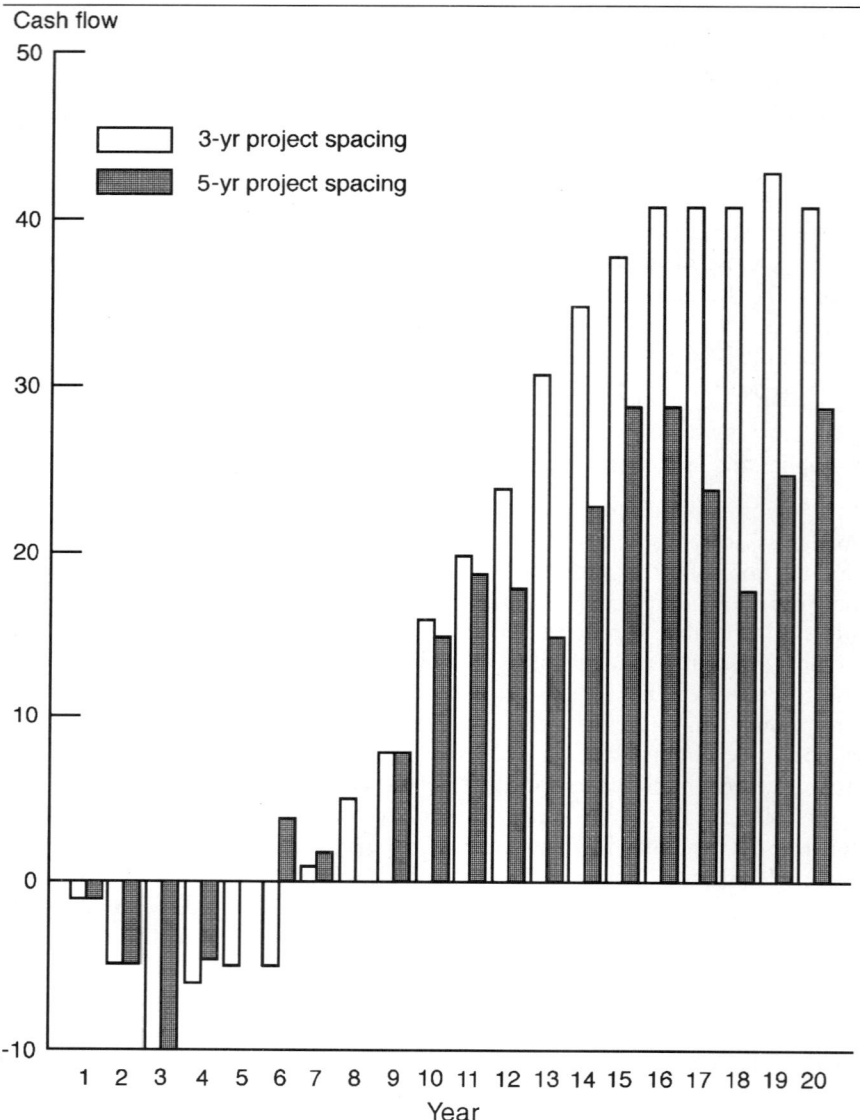

Figure 6.2 Effects of innovation project spacing
Source: DWB

Subsequent large injections of both cash and management, together with collaboration with a Japanese manufacturer, eventually restored it to health. The whole process took about twenty years, approximately the time scale of Figure 6.1.

A more cheerful tale is that of Glaxo which, in 1972, was the subject of an unwelcome takeover bid from Beecham. The bid was inspired by the belief of the then Chairman of Beecham, who was distinguished both as a businessman and as an economist, that only large companies would be able to afford sufficient R and D and other innovation expenditure to compete in the pharmaceutical industry of the future. Glaxo found the Beecham approach unwelcome and attempted to form a defensive merger with Boots, but both proposals were referred to the Monopolies Commission (now the Monopolies and Mergers Commission) which ruled against them both on the grounds that innovation would be inhibited by further concentration in the industry.

Thus left on its own, Glaxo, which had a long tradition of R and D and had been spending more on it than either Beecham or Boots before the bids, increased its concentration on R and D and expanded its overseas marketing. In accordance with this strategy, which was explained to shareholders in its annual reports, Glaxo made heavy investments which depressed its reported profits from 1977 to 1980. By 1979, however, it was able to report promising results from what was to become Zantac, eventually the biggest-selling drug in the world, and on which was based Glaxo's subsequent growth to become, first, the largest pharmaceutical company in the UK and, after the merger with Wellcome in early 1995, the largest in the world.

Although the innovation investment made by Glaxo depressed profits in the 1977–80 period, the company was careful to maintain its dividends, as can be seen from the figures shown in Table 6.1.

Disclosure of R and D expenditure was not required at the time, but Glaxo did report the rate at which it was spending, and the figures show, for example, that the decrease in profit between 1979 and 1980 was about equal to the increase in R and D expenditure between the two years.

Table 6.1 Glaxo Holdings plc: financial performance, 1974–83

	1974	*1975*	*1976*	*1977*	*1978*	*1979*	*1980*	*1981*	*1982*	*1983*
Profit pre-tax & int £M	49.1	49.2	81.4	95.1	94.8	80.1	75.1	99.6	148.5	205.7
ROCE %	29.2	25.0	30.6	29.2	28.6	21.3	18.8	22.5	30.9	36.4
R&D rate £M/y						32.5	38.0	45.0	53.0	70.0
Profit after tax on ord. shareholders' funds £M	22.0	20.3	34.7	41.6	41.5	47.5	41.8	60.6	80.4	115.4
Dividend per share p	2.5	2.9	3.5	3.9	4.3	5.7	6.8	8.1	10.0	12.9
Retentions £M	27.0	22.5	36.9	43.9	44.2	49.6	43.4	60.2	80.6	112.3

Source: Annual reports
Note: Zantac was introduced during the year ended 30 June 1983.

USES OF THE MODEL

The use of the cumulative cash flow curve for an innovation (Figure 4.2) as a tool for management of individual innovation projects was discussed in Chapter 4 with special reference to its use by Hewlett Packard. In this section, we look more deeply into its use in the context of strategic innovation management and as a component of the model of a company or business as a succession of innovations. This model, which contrasts with the accounting model of a company as a machine or 'black box' for processing money, usefully illuminates a number of strategic issues, and is helpful in pointing the way to developments which might integrate the innovation dimension into the financial management and reporting of companies.

How accurate is the model?

The cash flow curve of Figure 4.2 is often reproduced, usually as a concept without figures attached rather than in a numerical form for a real innovation. Although the curve would hardly have continued in use if it did not reasonably and usefully represent reality, it is worth while to spend a little time considering its validity.

First, the curve by definition starts at the origin and, if we remain with the definition of innovation as the successful exploitation of new ideas, it must be correct in ending up at a cash positive point. Success demands at least this.

Second, it is reasonable to expect that the exploitation phase will normally last longer than the development phase of a project. Unless the product is being developed for some specific short-term market, such as a major exhibition, sporting event, or similar selling opportunity which is highly concentrated in time, the risks of embarking on a long development project which is likely to be saleable for a comparatively short time are too great. In markets which demand both a reasonable degree of technical sophistication, requiring substantial development effort for each product, and are highly competitive or fashion-conscious, the result is to put a premium on shortening development time. The consumer electronics industry, to which Hewlett Packard belongs, and the automobile industry, in which product development times are a major concern of management, are notable examples of this combination.

Third, the curvatures at the various points of the cash-flow history reflect the realities of innovation. Expenditure in the early days of a project tends to be relatively small and to build up at an increasing rate for a time, giving the concave downwards shape to the initial part of the trajectory. Similarly, receipts tend to come in slowly at first and more quickly thereafter, producing the concave upwards shape of the next part of the curve. In the final stages, sales tend to drop away slowly at first and then more rapidly, giving a concave downwards shape once more.

Although the general shape of the curve is perhaps the one reasonably common feature of innovation across a wide range of types and activities, it cannot be expected that its precise shape will be the same, even from innovation to innovation within the same company or industry, and still less from industry to industry. A more technical complicating factor when comparing published curves is that the definition of what constitutes 'cash flow' may differ from source to source. Hewlett Packard uses profit as the measure of return, which is convenient and appropriate within one company but less so for general use or for comparisons between industries. As Chapter 2 has shown, profit is a somewhat slippery concept, highly dependent on conventions and judgement which may vary from case to case and time to time.

The pharmaceutical industry

The prescription pharmaceutical industry is the heaviest spender on R and D as a proportion of sales (CSO 1995) and probably on innovation, about which financial information is scarce. The industry funds all this expenditure itself, unlike some R and D-intensive industries which receive substantial public subsidy. It is, nevertheless, a very profitable industry and, in consequence, is subject to a good deal of public scrutiny since the buyers of its products are largely public bodies or large cost-conscious organisations of one type or another.

The pharmaceutical industry is therefore very aware of the need to produce good evidence that its innovation spending is both necessary and fruitful and it is, in consequence, much better documented than most others. As mentioned above, its products are quite distinct from one another; and its well-defined development process makes the gathering of data about innovation easier. Once a chemical entity has been found to have therapeutic effects, it must go through a series of stages to establish its safety and efficacy before it can be submitted for approval for marketing. Patent protection is invariably obtained at an early stage so that companies can reveal progress in development without compromising their commercial interests. Recognising that their future depends on the flow of new products, and that their shareholders have a legitimate interest in knowing what is in the pipeline, pharmaceutical companies routinely give reports on the progress of new products under development.

Grabowski (1991) has published a curve showing the annual cash flows from a typical pharmaceutical innovation. The data were based on a study of the costs and returns involved in the introduction of new chemical entities (NCEs) in the US in the 1970–79 period. The cumulative cash flow derived from Grabowski's curve is shown in Figure 6.3.

It can be seen that this curve is of the standard type. Its time span is about twice that of Figure 4.2, but the ratio of return to outlay is considerably worse. It is in fact about 1.8, compared with about 7 for the curve of Figure 4.2. The main reason for this discrepancy is that the return is measured by Grabowski

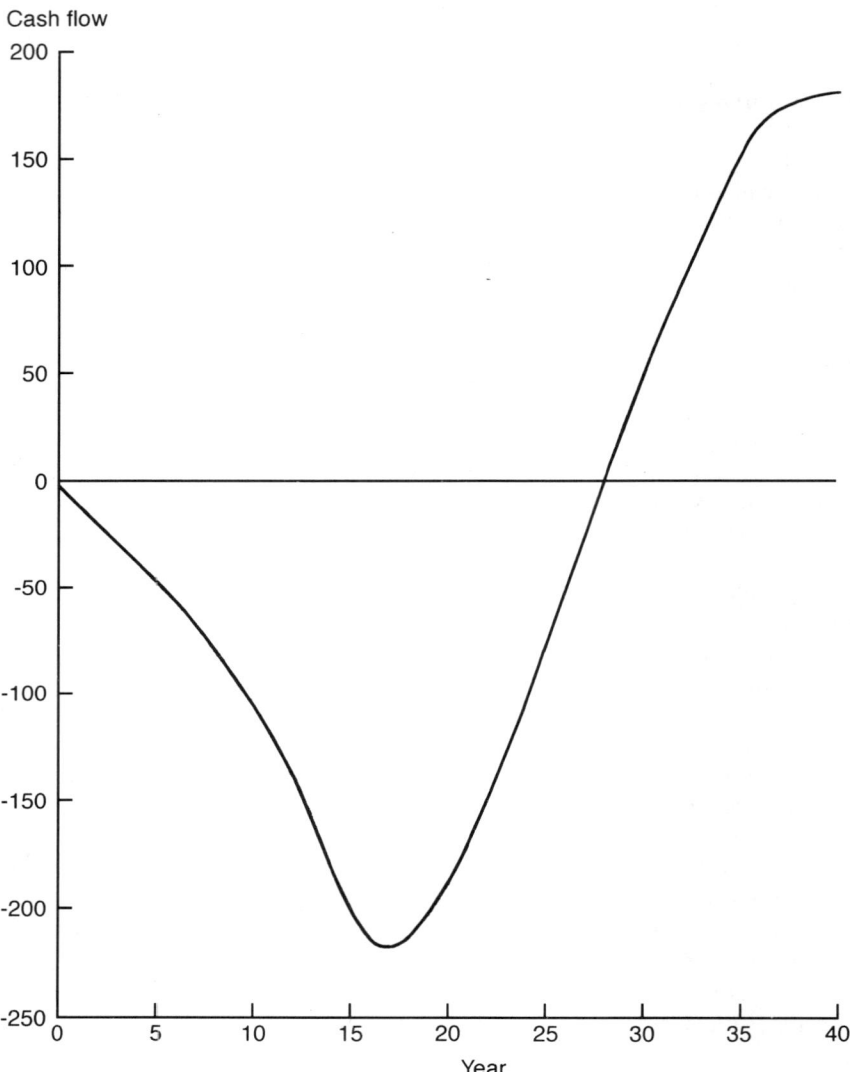

Figure 6.3 Pharmaceutical innovation cash flow
Source: DWB from Grabowski (1991)

(1991) as profit, estimated at 40 per cent of sales, less tax at 35 per cent, or 26 per cent of sales revenue, rather than as sales revenue or value added.

Grabowski and his colleagues have done extensive analysis of the economics of R and D in the pharmaceutical industry, calculating returns on the investment by means of DCF techniques. One of their results is that the returns on

R and D have been comparable to the opportunity cost of capital. They also find that the break-even time for pharmaceuticals is around seventeen years after marketing, based on discounted net present values of cumulative cash flow. According to Grabowski (1991), marketing would take place at about year fourteen. Figure 6.3, which is based on undiscounted values, shows break-even on this basis to be achieved at about year twenty-eight, fourteen years after marketing.

Implications for pharmaceutical companies

Using the cash flow curve of Figure 6.3, it is possible to produce a model for a pharmaceutical company on the same lines as that of Figure 6.1. This model predicts that a new pharmaceutical company would take about thirty to thirty-five years to reach a reasonably stable financial state. This prediction seems to be borne out by the experience of ICI Pharmaceuticals (now Zeneca) which traces its origin to a Board resolution in 1936 which made a grant of £15,000 per annum (about £500,000 in 1995) for an initial period of five years to fund research in the Dyestuffs Group on synthetic organic pharmaceuticals with a view to finding new remedies and replacing imports. The activity was successful, but was not financially self-supporting until the early 1960s, some twenty-five years later (Newbould 1995).

The development of pharmaceuticals has become a longer and more expensive process than when ICI first entered the field. A more recent entrant which has still to become profitable is British Biotech plc (originally known as British Biotechnology), which was founded in 1986. The company described itself in its 1995 annual report as a development stage pharmaceutical company which had no product revenues as yet and a prospect of absorbing cash until products were commercialised. Its admittedly ambitious goal was to build an international pharmaceutical company on the basis of a flow of breakthrough products from R and D. It believed, in common with this book, that this goal could be achieved only on the basis of a broad pipeline of products at different stages of development. By 1995, British Biotech had five products in clinical trials and several more in discovery and pre-clinical research.

British Biotech's strategy is to target cancer on the grounds that the opportunities are good, especially given the company's own strengths, and that the relative shortness of the clinical trials and the modesty of the marketing requirements in this field of acute diseases bring it within the financial scope of a new company.

After its launch in 1986, the company raised further finance in 1988, 1989, and 1991. It achieved a listing on the London Stock Exchange and on the New York NASDAQ exchange in 1993, and raised further funds through a rights issue in 1994. Warrants were issued entitling the recipients to buy further shares in 1996. At about this time it also sold a profitable reagents and diagnostic business which it had developed, in order to concentrate on its main target. In 1994, a subsidiary company was formed to pursue the application of some products outside the field of cancer.

The company has received modest amounts of money, shown as turnover in its accounts, from European and other sources, either by way of grants or as contributions to shared programmes, but the bulk of its R and D has been funded from its equity. The main features of its financial progress are shown in Table 6.2. At the end of its financial year on 30 April 1995 it was still in Stage 1 of Figure 6.1, with its increasing losses having reached a total of £85 million.

British Biotech satisfies the criterion of Chapter 5 in having a limited number of committed shareholders, with its Annual Report for 1995 showing that 38 per cent of its shares were held by only five shareholders, the largest of which held 13.8 per cent.

The pharmaceutical portfolio

The major purchasers of the products of the pharmaceutical industry, such as the National Health Service (NHS) in the UK, are very aware of the costs involved and exercise considerable pressure to keep them under control.

One response of the companies to increasing price pressure has been to make strenuous attempts to reduce the development times of drugs, by minimising the gaps between essential testing stages or even overlapping them, accepting the risk that bad late results from an earlier stage may render any work already done on the later stages a complete waste of money. Another response has been to be more ruthless in selection at the early stages and to reduce the number of compounds subjected to later-stage testing.

Using the extensive world-wide data on pharmaceutical R and D and its products, Drews (1995) has given some figures for the cost elements in the research-based pharmaceutical industry as follows:

(a) Discoveries taken to development: 40 per cent
(b) Development projects taken to market: 10 per cent
(c) Annual cost per discovery: $17 million
(d) Discovery project length: 4 years
(e) Development time: 6 years
(f) Annual sales per product: $400 million
(g) Average product life span: 17 years

Drews estimates that a major company will need to introduce one new product per year which, it will be remembered, is the avowed target of Zeneca. There are twenty such companies, giving rise to a need for 500 discoveries (ie 20/(0.4 x 0.1)) per year, or 2,000 projects in being at any one time. The annual discovery cost will be $8.5 billion, and the total R and D cost will be several times as great, depending on the ratio of development to research costs. Drews estimates this to vary between 2.3 and 4, making a total required annual

R and D cost in the range $28–40 billion compared with the current rate of about $17 billion. He therefore suggests that the current innovative effort cannot support an industry of the present size and that some shrinkage will occur unless the efficiency of development is improved or biotechnology comes to the rescue by producing new discoveries more cheaply than by the conventional methods.

Table 6.2 British Biotech plc: seven-year summary, 1989–95

	1989 £000	1990 £000	1991 £000	1992 £000	1993 £000	1994 £000	1995 £000
Turnover:							
Continuing operations	2,655	2,349	2,934	1,220	2,519	2,318	3,191
Discontinued operations	646	1,402	2,307	3,785	5,328	1,668	—
Total	3,301	3,751	5,241	5,005	7,847	3,986	3,191
Operating (loss)/profit							
Continuing operations	(1,908)	(4,563)	(7,941)	(15,171)	(17,711)	(23,224)	(29,052)
Discontinued operations	(264)	63	79	259	409	181	—
Total	(2,172)	(4,500)	(7,862)	(14,912)	(17,302)	(23,043)	(29,052)
Loss on disposal						(142)	
Net interest received	245	2,798	1,736	3,300	4,206	1,662	2,721
Loss before tax	(1,927)	(1,702)	(6,126)	(11,612)	(13,096)	(21,523)	(26,331)
R and D expenditure	1,908*	5,044	8,676	14,116	16,748	20,650	26,210
Assets employed							
Fixed assets	6,349	9,217	13,753	14,595	16,993	18,671	23,131
Net current assets	4,057	21,112	10,634	36,006	47,952	24,756	39,683
Long term debt	(4,214)	(4,272)	(4,456)	(4,245)	(4,064)	(4,028)	(3,727)
Net assets	(6,192)	26,057	19,931	46,356	60,881	39,399	59,087
Capital employed							
Share capital	439	876	876	1,433	1,809	1,811	2,418
Share premium account	10,008	31,138	31,138	68,618	95,863	95,907	137,995
Warrants							3,337
Profit and loss account	(4,255)	(5,957)	(12,083)	(23,695)	(36,791)	(58,319)	(84,663)
Shareholders' funds	6,192	26,057	19,931	46,356	60,881	39,399	59,087

Source: Annual reports 1993, 1994, 1995
*In the summary given in the 1993 report, the term R and D was used in place of operating loss from continuing operations. In the 1994 and 1995 reports, the two are distinguished. The figures quoted above, except that for 1989, are from the 1994 and 1995 reports. Where the series overlap, those quoted in 1993 were slightly lower.

INCORPORATING THE MODEL INTO COMPANY ACCOUNTS

Drews (1995) assumes that the current ratio of R and D expenditure to sales in the pharmaceutical industry will be maintained. Clearly, there is a limit to how much can be spent on R and D, given the other demands on a company's budget, not least the demands of other aspects of innovation, and the board has to decide what this limit should be. If innovation is as central to the performance of a company as it clearly is in the pharmaceuticals case, some method of integrating the expenditure on it into the company's accounts is desirable. Although the financial demands of R and D are less onerous in other sectors, the arguments put forward in Chapter 1 suggest that the strategic importance of innovation is no less and it may well be that the lower costs of R and D are made up by increased costs of other parts of the innovation process.

The financial model of a company as a series of innovations, shown in Figure 6.1, provides a basis for integrating innovation expenditure into the company's accounts. In Stage 3 of the company's development, for years sixteen to twenty-five, it is spending just enough on innovation to replace old products with new ones, and is therefore running at a constant size. It is necessary to maintain this level of innovation expenditure if the company is to preserve its earning capacity into the future. Remembering the definition of profit as the maximum which can be taken out of the company while leaving it as well off as it was before, this innovation expenditure must be considered as not properly part of the profit. Spend less on innovation and the company eventually shrinks, as Stages 4 and 5 of the model clearly show: spend more and the company grows, as in Stage 2. In the first case, the company is worse off at the end of the period, while in the second it is better off.

The case is, in some respects, analogous to the depreciation provision. Theoretically, at least, the depreciation provision preserves the total value of the business by transferring into cash the diminution in the value of the fixed assets which has occurred through wear and tear or other factors during the year, or other period for which the accounts are made up. A company can spend more than the depreciation provision on acquiring new assets, in which case it will increase its asset base; or it may spend less, in which case its asset base will decrease. In either case, the change in the fixed asset base would be offset by the change in the current assets. The analogy with innovation breaks down at this point because the depreciation provision relates to an asset which appears on the balance sheet, and the provision is therefore legally protected, while innovation expenditure relates to an intellectual asset which is not capitalised. This expenditure cannot therefore be legally protected and could, in consequence, be (improperly, according to the current argument) counted as profit by a board of directors with a short-termist orientation.

Assuming that the board is properly conscious of the economic definition of profit and persuaded by the model, the question arises as to what the innovation expenditure actually is in the stable state of Stage 3. It can be shown (Budworth 1987) that the ratio of return to outlay for the average innovation, as shown in Figure 4.2, is the same as the ratio of inward cash flow to innovation expenditure for the multi-project company as a whole when it is in the stable state. This relationship provides a method of relating the financial characteristics of the business as a whole to those of the average innovation.

It was mentioned above that Grabowski's cumulative cash flow curve for a pharmaceutical innovation (Figure 6.3) took profit as its measure of return. Because of the subjectivity of profit measurement, and its dependence on a particular pattern of costs and receipts, it is preferable for a general model to measure return as value added, the difference between sales and purchases. As was pointed out in Chapter 2, value added is both easy and unambiguous to measure and it represents the resources which management has at its disposal inside the business. An alternative would be to use sales income as the measure of return, but sales are not obtained without some external expenditure in most cases, and it is the difference which matters.

To accord with accounting conventions, innovation expenditure in the model is taken as the revenue component of the total expenditure, excluding the capital items which would appear on the balance sheet. It is defined as revenue expenditure which is made in the expectation that it will contribute to the profits of future accounting periods, rather than of that in which it is incurred. A substantial part of it will usually be employee costs, but there will be contributions from most, if not all, the other headings on the profit and loss account. Although at the internal, management accounting, level the innovation content of each of these items would have to be separated, the more important strategic figures are the total innovation expenditure and its ratio to the value added. This 'innovation ratio' is the analogue for intellectual capital of the depreciation ratio on fixed assets.

A single figure for the innovation expenditure could be incorporated into only the 'functional' form of the profit and loss account (Table 2.3), in which the depreciation provision does not appear. From a strategic point of view it is desirable to have both in view simultaneously since both represent non-trading aspects of the company's financial affairs. This end can be achieved by incorporating innovation expenditure in a value added statement, in a development of the way in which Amersham once presented its R and D expenditure (Table 2.12).

Application to mechanical engineering and pharmaceuticals sectors

The curve of Figure 4.2 has an innovation ratio of 0.14, a figure which is reasonably representative of the mechanical engineering sector. Using figures for the pattern of costs for this sector based on Cox and Kriegbaum (1989) and other sources, the value added statement for a mechanical engineering company would be approximately as shown in Table 6.3.

Table 6.3 Distribution of value added for a mechanical engineering company

Sales	1000
External purchases	(500)
Value added	500
Employment costs (net)	(320)
Interest	(10)
Tax	(25)
Depreciation	(25)
Innovation	(70)
Retentions and dividend	50

Source: DWB

Such a company would probably have a capital employed of around 400 units, of which 300 would be shareholders funds, which is consistent with the depreciation provision of 25 units, and 100 would be loans, accounting for the 10 units of interest. Tax at the current corporation tax rate of 33 per cent is charged on the profit (before tax) of 75 units.

The key advantage of exhibiting innovation expenditure in this explicit way is that it emphasises its vulnerability. Marginal increases in the two large deductions from the income – the external purchases and the employment costs – can soon bite into the innovation expenditure, given that the interest and depreciation charges are more or less fixed, and that there will be demands from shareholders for profit and hence a tax charge.

The pattern of costs would be quite different in different industries. For the pharmaceutical industry, using Grabowski's (1991) figures and some other information, the distribution of value added would be approximately as shown in Table 6.4.

It can be calculated that the innovation ratio, on the same basis as that for mechanical engineering, is about 19 per cent, as opposed to 14 per cent. The difference between the industries thus appears to be not so much in the innovation intensity in terms of value added, but in the value added as a percentage of sales and the share of employment costs for production in the value added.

Table 6.4 Distribution of value added for a pharmaceutical company

Sales	1000
External purchases	(260)
Value added	740
Employment costs (net)	(150)
Interest	(0)
Tax	(140)
Depreciation	(50)
Innovation	(140)
Retentions and dividend	260

Source: DWB

OTHER IMPLICATIONS OF THE MODEL

The need for a minimum number of innovation projects

In deriving the innovation ratio we made use of the model in its equilibrium state, in which the innovation process produces new products at a rate just sufficient to replace the old ones which have reached the end of their market lives. To achieve this equilibrium state, new projects were started at three-year intervals. With the cash flow pattern for an innovation used for the model (Figures 4.2 and 5.1), the company in this dynamic equilibrium state has about seven projects running at any one time, each project being at a different stage of its life. The exact number and spacing of projects needed to achieve reasonable equilibrium depends on the exact shape of the cash flow pattern but, remembering that it must be of the same general shape as that assumed, for the reasons given above, the general conclusion is valid. For the somewhat different Grabowski cash flow pattern of Figure 6.3, for example, reasonable stability can be reached by starting new projects at five-year intervals which results in there being about eight such projects current at any one time.

For a real company, the picture would be somewhat more complicated. As it is completed projects which matter, since it is only from these that a return is obtained, the cash flow pattern of the average innovation used in the model must include allowance for those projects which prove to be unsuccessful at some stage in their development and are abandoned. The figures given by Drews provide guidance on what attrition factors to adopt for the pharmaceutical industry, and his target of a new product introduction every year for a major company is more demanding than the five-year sequence which satisfies the idealised minimum condition.

There are other reasons for thinking that, in practice, a company would need to have more innovation projects than the minimum needed for overall financial stability. If we look at the individual contributions to this stability made by the innovation expenditure on the one side of the profit and loss account and the value added by sales on the other, we find that both are varying quite markedly from year to year, with their variations offsetting each other. This effect shows up very clearly if we calculate the innovation ratio for each year. Figure 6.4 shows that the innovation ratio, even in the stable period, varies over the three-year cycle to an extent which would make management extremely difficult and lead to variations in profit and capacity to pay dividends which shareholders would find unacceptable. If the frequency of projects is increased to one per year (Figure 6.5), then the innovation ratio remains constant in the stable period, making management much easier.

Whatever the details, however, the conclusion remains that a business needs to have a series of innovative projects spaced out in time if it is to be reasonably stable in cash terms.

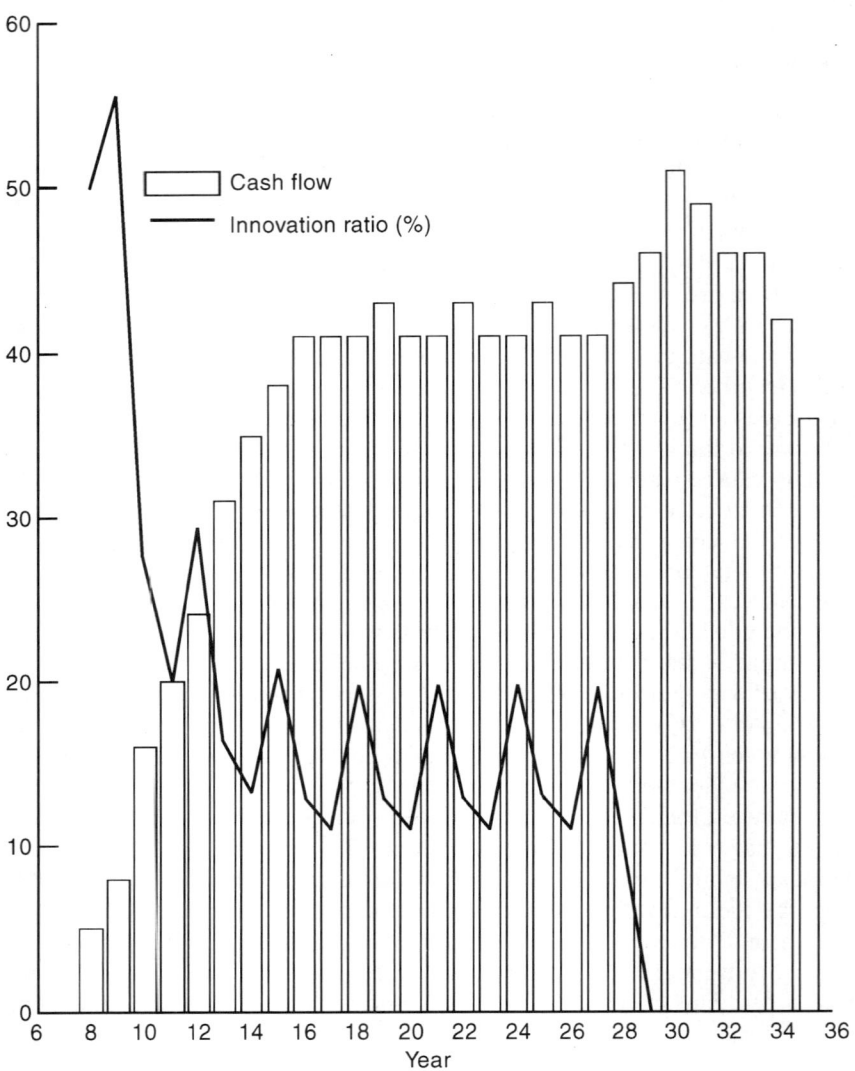

Figure 6.4 Projects at 3-year intervals: cash flow and innovation ratio
Source: DWB

This conclusion is of the utmost importance for the strategy of a company for, as has been emphasised at various points in this book, cash is the key to success. Lack of cash is what causes companies to fail and irregular cash flow leads to many problems. If a run of success enables a company to build up a generous cash balance which the directors might wish to reserve for use in

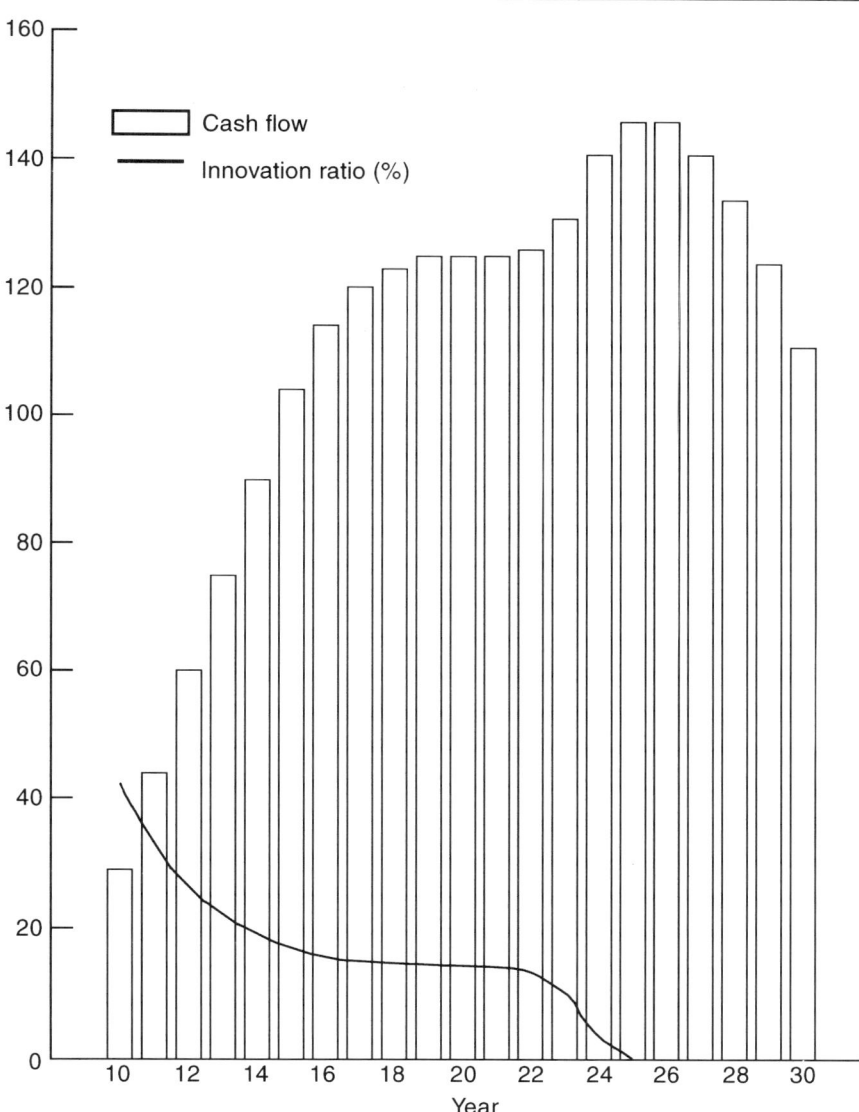

Figure 6.5 Projects at 1-year intervals: cash flow and innovation ratio
Source: DWB

harder times, there will be demands from shareholders to distribute it or to use it for acquisitions since the returns will be higher in the short run in either case. The Chrysler car company in the US had a problem of this kind in early 1995. The UK metrology equipment company Renishaw has a policy of maintaining a large cash reserve, of around half a year's turnover, and can maintain this policy because the two founding directors control just over 50 per cent of the shares.

The more usual problem, however, is that of shortage of cash. The size of an innovation project is to some extent fixed by the characteristics of the industry in which it occurs, through a combination of technical, regulatory, and competitive factors, and is beyond the control of an individual company. For example, the ABPI (1993) quotes a figure of £200 million for a new pharmaceutical, up from £50 million in 1985, while the implication of the figures given by Drews, quoted above, is that it is in the range $1,400–2,000 million. The cost of developing a new car, aeroengine, or aircraft is in the same range, with figures of around $1,000 million being quoted for cars (Lutz 1994).

A company may therefore be too small to be able to afford to have the required number of innovation projects in being at any one time in which case it will have an unstable cash flow with severe shortages when it is in the expensive phases of developing a new product, and only an adequate flow when the product brings in a return. (If the cash flow was more than adequate at this stage, the company could expand its innovative activities with the aim of reaching a size at which it would be more stable.)

The problems of the undersized company

A company in the position of being too small to afford a sufficient number of innovation projects to maintain financial stability can choose one or more of several possibilities for coping with the problem. It can attempt to live with the situation, accepting that it will have a somewhat precarious and erratic life; it can merge with another company in the same industry to form a more viable unit; it can become a subsidiary of a large parent which can cope with cash flow irregularities by diluting them in a larger pool; it can acquire another business which is in cash terms complementary; or it can seek help from government.

Living with the situation is possible only for a company which is not exposed to stock market pressures. As the example of British Biotech shows, the stock market can, in some circumstances, give strong support to an immature company which shows promise of eventual success, and will tolerate losses for an extended period in the hope of major gains. It will not, however, be very happy with a company which is clearly too small for the business that it is in, and will bring pressure to bear for it to take one of the other possible courses. A private company can survive, usually by building up reserves and then using them for developing a new product. Examples can be found in very small car companies and in the scientific instrument industry. Such companies are, however, intrinsically precarious and are unlikely to survive the departure of the founders.

Examples in the pharmaceutical industry

Examples from the pharmaceutical industry have been discussed above. Until the Glaxo Wellcome merger in early 1995, no company had more than about 4 per cent of the world market for pharmaceuticals (Blaker 1994), and the industry remains well dispersed, with over fifty significantly active companies. As we have seen, however, the cost pressures on it arising from a combination of more rigorous testing and more demanding customers have already led to the departure of two UK research-based companies which were too small to be viable. The projections of Drews suggest that more will follow. Mergers, as in the Glaxo Wellcome case, and diversification into other parts of the health care industry, as in the case of SmithKline Beecham, which has taken over a health care management organisation in the US, are being pursued as strategies. It is notable that the larger companies are responding to what they see as new strategic imperatives.

Seeking shelter in the arms of a large conglomerate seems to be an unlikely response of a small pharmaceutical company to current strategic pressures. The unfortunate experience of the Distillers company, which diversified into pharmaceuticals with thalidomide, stands as a warning that the industry is not one for non-specialists.

The pharmaceutical industry has close relations with governments, who are its major customers in many countries, but has kept these relationships on a customer-supplier basis. Rather than seek government subsidy for its innovation activities, it has sought to achieve and maintain recognition that these activities are a necessary part of staying in business – the main theme of this book – and that the costs incurred have to be recouped through sales. In the UK, there is a voluntary agreement – the Pharmaceutical Price Regulation Scheme (PPRS) – between the NHS and the ABPI which defines its purposes as being to:

1 secure the provision of safe and effective medicines for the NHS at reasonable prices;
2 promote a strong and profitable pharmaceutical industry in the UK capable of such sustained research and development expenditure as should lead to the future availability of new and improved medicines; and
3 encourage in the UK the efficient and competitive development and supply of medicines to pharmaceutical markets in this and other countries.

The PPRS applies only to sales to the NHS, which amount to around £2.5 billion. Companies which sell more than £20 million to the NHS, of which there are thirty-five, accounting for about 86 per cent of total sales, make annual returns on which individual negotiations are based. A tolerance of plus or minus 25 per cent on the target profit rate for each company is allowed. The version of the PPRS which is due to run for five years from October 1993 permits companies to make a return on capital employed of 17–21 per cent, unless their turnover is more than 3.75 times their capital employed. Such companies, which are normally those mainly engaged in distribution, are given a permitted return on sales. Normally, government accounting conventions do not permit the inclusion of intangible assets in capital employed, but the PPRS allows for negotiation about the inclusion of goodwill, trade marks, and patents.

Allowable promotion costs are defined by a formula which combines a flat rate element of 6 per cent and a sales-related element. It is intended to limit the overall total to 9 per cent of sales, a figure considered by the ABPI to be modest by the standards of the industry world-wide. This figure is one of the few reliable indicators of the costs of the non-R and D parts of innovation and again it consists largely (about two-thirds) of employment costs (ABPI 1993).

Allowable R and D costs under the PPRS are negotiated individually, taking into account the average spend for the UK industry, the pattern of the company's investment in the UK, and the company's world-wide reference level of R and D expenditure. The final figure is expressed as percentage of sales.

There is also a provision that transfers of raw materials, intermediates, or finished goods from an affiliated company should be made at 'arm's length' prices, thus raising the issue of valuation.

The motor vehicle industry

World-wide, the motor vehicle industry is a good deal further towards maturity than the pharmaceutical industry. In its early days, about a hundred years ago, it was the province of a few enthusiasts who rapidly explored almost all the design possibilities until the definitive form of the front-engined, four-wheeled, rear-driven, configuration emerged in about 1904. For a period, the design was sufficiently standardised and the production methods so simple that almost any well-equipped mechanical engineering workshop could make a car with a little help from specialists. Many of them did, and the number of manufacturers proliferated until the depression of the 1930s reduced their number drastically.

In the subsequent period, manufacturing methods became more specialised, reducing unit costs provided that development and capital costs could be shared over a larger output. Technical developments enabled one of the early rejected design configurations – front-wheel drive – to become practicable, widening the range of design types on offer. Further manufacturing concentration took place aided by increasing trade between manufacturing countries, and safety regulations began to make development even more costly. The number of manufacturers dropped even further, and the minimum viable size rose once more to a figure which is probably around 500,000 units per year for a mass producer.

This history is of interest, not just in itself, but because it is typical of the development of an industry based on a new product. There is an early period of experimentation, when many different designs are explored, followed by a period in which a dominant design emerges. Competition then focuses for a time on adding features to this basic design, with the market gradually widening from the early enthusiasts to a more general public. Finally, the product becomes something of a staple or commodity, when competition focuses more on price or value for money. The personal computer, which has developed very rapidly, has gone through most of this cycle in little more than ten years. The much more expensive nuclear power, on the other hand, has reached only the dominant design stage after some fifty years of development.

More formally, this evolutionary process has been called the standard product cycle by theorists of technological change, who distinguish three phases of development once a new technology has become established and embodied in a new set of emergent products. The first phase, introduction, is characterised by major product innovation involving rapid technical advance, labour intensive production methods applied to low volumes of output, and competition through product performance. In the second phase, growth, process innovations begin to dominate, production becomes more capital intensive, and competition is on the basis of the performance and quality of a decreasing range of products. In the final, mature, phase, technical change focuses on incremental improvements designed to reduce product cost, with price as the chief area of competition.

The specialist luxury car maker Jaguar provides a good example of size sub-viability which ended in being taken over by the vastly larger multinational Ford. Jaguar was the creation of Sir William Lyons and flourished from the late 1930s until Lyons retired in the mid-1960s, entrusting his already-vulnerable company to the larger, but ultimately disastrously unsuccessful, British Leyland. Jaguar was floated on the stock exchange as an independent company once more in 1984. Its accounts for the 1986 calendar year clearly showed that the cost of developing a new model (the XJ6), had put a great strain on the company's resources (Budworth 1988). At this time the company was apparently doing well with a high share price, but its strained cash position, which showed up in its Z-score despite the presence of net cash of £65 million in its 1988 accounts (Taffler 1995), made it vulnerable to recession, and it was taken over by Ford in 1989. Ford paid £1.86 billion for Jaguar which was reported to have lost a further £776 million up to 1993.

Rolls-Royce Motor Cars is also a very small company for its industry. Although Rolls-Royce competes in a very special niche for luxury vehicles of high prestige, it still has the problems of replacing products from time to time. Its position as the largest single part of the engineering group Vickers offers it some protection but, as it constitutes around 30 per cent of the group, that protection is limited. The automotive engineering activities of Vickers also include a specialist maker of racing engines, Cosworth, and some other activities, so that Rolls-Royce is not separately distinguished in the Vickers accounts.

Rolls-Royce Motor Cars was very badly hit by a downturn in orders in the recession of the early 1990s (Table 6.5).

Table 6.5 Rolls-Royce Motor Cars: sales 1990–94

Year	1990	1991	1992	1993	1994
Sales (units)	3,333	1,722	1,378	1,360	1,414

Source: Annual reports of Vickers plc

The company responded, first, by improving its production methods to lower the break-even production figure by around 50 per cent to about 1,300 and, second, by developing new products by evolution from those already in production. Looking further ahead, the company decided to introduce completely new models on an eight to ten year cycle, as opposed to its previous cycle of fifteen to twenty years. There were limits on how far it could go in this direction, however, because, regardless of the availability of funds for developing new models, buyers of Rolls-Royce cars do not expect them to date quickly so that too frequent a model change could be counterproductive.

One major step towards funding the extra innovation costs which the shortening of model lives would involve was to make an agreement with BMW to supply engines for future models.

The aerospace industry

There are at least twenty major pharmaceutical companies world-wide, together accounting for about 67 per cent of the market in 1992 (Rickwood 1993), and a comparable number of major automobile companies, but the concentration in aerospace is much higher. There are only three commercially-competitive makers of large aeroengines – GE, and Pratt and Whitney in the US, and Rolls-Royce in the UK – and three commercially-competitive makers of large civil aircraft – Boeing and McDonnell Douglas in the US and the European consortium Airbus Industrie. Further, the development costs of a new aircraft or aeroengine are comparable with those of an automobile at around £500–1,000 million, while the time scales are comparable with those of the pharmaceutical industry. For aircraft, the investment period is around five years, with break-even at ten or more years later; while for aeroengines these periods are approximately doubled (HL 1993). The problems of coping with these costs and time scales are therefore very marked and the companies in the aerospace industry find themselves in a difficult position. This difficulty has arisen from the unique historic relationship between finance and innovation in aerospace, in which the demands of innovation have largely outweighed those of finance.

Because of the intimate connection with defence, the aerospace industry traditionally has close connections with governments. Defence product development is largely financed by governments, and some of the knowledge thus gained can be used by the companies for civil applications, but there is also a tradition of government support for civil aerospace. In the UK this takes the form of 'launch aid' in which the government provides an agreed sum towards the development and launching costs of an aircraft and recoups its outlay by a levy on sales.

The reason why launch aid is needed was explained very clearly in the Chairman's Statement in the annual report of British Aerospace plc for the year ending on 31 December 1986:

> We must ensure that as a company in the private sector, we have a balance of business activities so that we are not overdependent on any one sector. In civil aircraft we have moved quickly to establish a position in the market place both with our own products and through Airbus Industrie. Although the company could fund the costs of a large new programme, such as the [Airbus] A330/A340, nevertheless we must pay due regard to the additional risk that this would add to the already significant level of funds employed on other existing civil projects. Without an acceptable level of risk participation outside the company, it would not be prudent to commit ourselves to these programmes, however deserving.

Or, in other words, the company was too small to carry the risk.

The report also explains launching costs and how they are dealt with in the accounts:

> The costs of launching a new civil aircraft project fall into three principal categories: design and development, jigs and tools, and education.
>
> Design and development expenditure arises mainly in the early years of a project and, with the exception of the A320 programme, is written off as incurred under the heading of launching costs in the profit and loss account. Initial launching costs on the A320 programme are covered by HM Government launch aid and, as a result, the design and development expenditure is carried forward and will be amortised, based on an appropriate assessment of sales, when deliveries commence.
>
> Expenditure on jigs and tools is normally incurred only after the decision to manufacture has been taken.
>
> Education costs arise early in the production cycle of a new aircraft when project skills and expertise are being acquired and unit costs are therefore relatively high. They comprise the excess of such unit production costs over the level of unit costs anticipated at a more advanced stage of the production programme.
>
> Expenditure on jigs, tools and education is carried forward in stocks for amortisation in respect of each project by reference to an appropriate

assessment of sales, subject to the status of the project and the overriding concept of financial prudence. The amortisation of jigs, tools and education is charged in arriving at trading profit. (From 1992, British Aerospace changed its accounting policy for jigs and tools, thereafter capitalising them into fixed assets and depreciating them over their useful life.)

Government aid is not the only means employed by the aerospace industry to counteract the problems of being too small. The collaboration through Airbus Industrie was a major development in the European aerospace industry, bringing together for specific projects British Aerospace and the principal French, German, and Spanish constructors.

Legally, Airbus Industrie is not a separate limited company, but a French Groupement d'Intérêt Economique (GIE). The GIE was at the time of the formation of Airbus in 1970 a feature only of French law, but was introduced on a EU scale under an EC Regulation of 1985 under the title of European Economic Interest Grouping (EEIG). A GIE or EEIG is a corporate body formed by contract between a number of parties in order to develop or facilitate the economic activities of its members and to improve or increase the results of those activities, not to make profits for itself. Consequently, it does not publish accounts and its results appear only implicitly in the results of its members.

Airbus Industrie has been active for over twenty years, and provides a good example of the build-up of a business (Figure 6.6). It is already the major global competitor to Boeing and, from time to time, there is talk of converting it into a limited company. It seems inevitable that sooner or later Airbus Industrie will absorb its members, forming one major international company to compete with the US manufacturers.

Aeroengines

The two US-based world aeroengine manufacturers are both subsidiaries of larger corporations, while Rolls-Royce has been independent since privatisation in 1987. In 1994, General Electric's aircraft engine business accounted for about 10 per cent of its turnover and operating profit, while in 1993 Pratt and Whitney and other aerospace interests produced about 25 per cent of the turnover and 12 per cent of the operating profit of United Technologies. (It made a loss in 1992.) In contrast, Rolls-Royce aeroengines accounted for around 60 per cent of its 1994 turnover and 35 per cent of its profit. The rest of Rolls-Royce's business is made up of its industrial power operations which were acquired in 1989 in order to lessen its dependence on aerospace and make more extensive use of its technological expertise, in accordance with the strategy of complementary acquisition.

These figures suggest that aeroengine manufacture is not a particularly profitable business, and indeed Rolls-Royce provides the classic case study of

Order

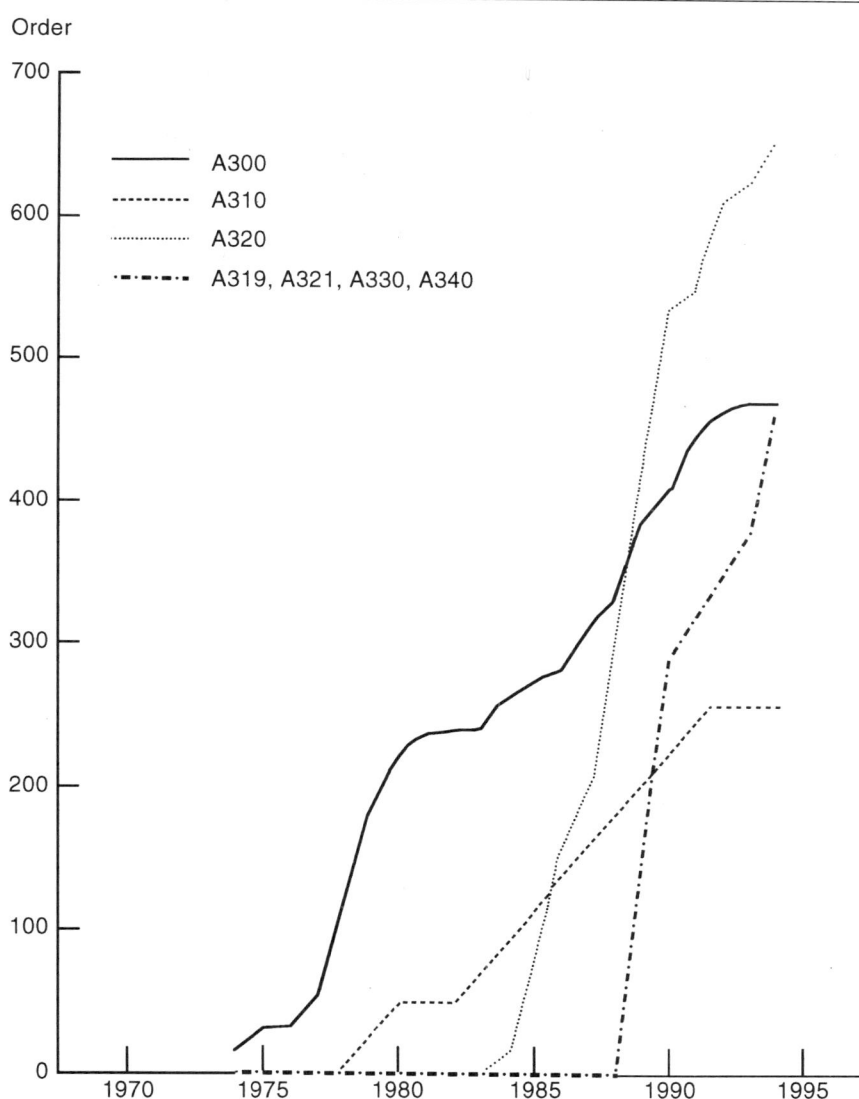

Figure 6.6 Airbus – cumulative orders by model
Source: Airbus Industrie

a company which failed, in 1971, because of heavy development costs. Because of its strategic and economic significance the company was taken into public ownership but its failure triggered an investigation under the Companies Act 1948 whose report (DTI 1973) gives a good account of what happened.

The development in question was the RB211 which marked a major technical departure from previous engine designs by having three shafts instead of two. This change promised greater efficiency and other benefits by enabling different parts of the engine to rotate at their individual optimum speeds rather than at a compromise speed. The engine was also designed to produce a thrust of about twice that of any previous Rolls-Royce product. This bold technological leap was regarded by Rolls-Royce as strategically necessary if it was to stay in the aeroengine business beyond the 1970s. In the light of subsequent events, this judgement seems to have been entirely sound, but Rolls-Royce underestimated the costs of development which, despite launch aid of 70 per cent of the original estimated costs, eventually overwhelmed the company.

One of the contributory causes of the underestimation unearthed by the enquiry was that the technical step represented by the RB211 was larger than it might have been following the abandonment of the development of an earlier engine, the RB178. Rolls-Royce had itself seen the need for the development of this engine and of a more general technology demonstration programme in order to minimise the height of the development steps to a level at which they could be planned with more certainty, but could not secure the government support which it needed to carry out such a programme.

In explaining the situation, which was already looking grim, to shareholders at the annual meeting at which the results for 1969 were presented, the Chairman of Rolls-Royce gave some useful background. He pointed out that the first civil gas turbine engine, the Dart, had sold for £7,000 in 1952 (equivalent to about £12,000 in 1969). This engine was still selling and also generating good spares business. One of the more recent engines, the Spey, originally sold for £65,000, whereas the initial price of the RB211 was expected to be in excess of £250,000. As the development costs varied roughly in line with price, it was clear that the development of new engines could not be financed from the sale of existing ones. Nonetheless, both to stay in business and to make use of fixed assets which were of no value for anything else (and whose value was no doubt having to be depreciated year by year at the expense of the profit and loss account), development of new engines had to be undertaken.

On the other side of the picture, the Chairman pointed out that the RB211 was one of a new breed of engines which represented something of a technological plateau and although it was expected to contribute only 13 per cent of turnover over the next five years, business from it could be expected for very many years to come. This prediction has again turned out to be entirely correct.

The inspectors found that by the early 1960s, Rolls-Royce (although it still had its car business at the time) had become largely dependent on the high-risk business of aeroengine manufacture, but believed that its future lay in this area, recognising its consequent dependence on the government both as customer and supplier of finance. It had given little serious consideration to major diversification, but concentrated on breaking into the major market for its type of product, the US. An approach from Lockheed to supply an engine for its proposed L.1011 had provided that opportunity, with the RB211.

One probable reason why the company gave little consideration to diversification was that its financial position had deteriorated from the early 1960s and it had, in consequence, become about 45 per cent dependent on borrowings for its capital. As early as its 1961 accounts, when there was no accounting standard for R and D, Rolls-Royce began to capitalise development expenditure, thus converting what would have been a reported loss of £1.7 million in that year into a reported profit of £2.5 million. This change was specifically made with a view to supporting the dividend, but although it provided the apparent profit for the purpose, it did not provide the cash out of which the payments had to be made. Profitability remained at a low level throughout the 1960s, and was not expected to recover until the early 1970s. In 1966, the then Chairman of Rolls-Royce listed the company's problems, in a letter to the Ministry of Aviation, as:

- The magnitude of the risk involved in launching civil engine projects, as a proportion of the total business and of its profits.
- Providing the finance.
- Dealing with the impact on profit of the annual spend on launching costs.

In terms of the model of a company as a series of innovations, Rolls-Royce was clearly still in the start-up phase, essentially of a new business, at the time of its collapse, even to the extent of a heavy dependence on borrowings. To survive this phase, as we have seen in Chapter 5, a company needs a restricted number of committed shareholders willing to sit out a lengthy period of losses in the interests of building up a viable long-term business. Not unreasonably, in view of its long history of success since its foundation in 1906, neither the company nor its shareholders saw it in this light and, given the long history of government support both for itself and its competitors, saw the continuation of that support as the way out of its difficulties. Another contributory factor was that Rolls-Royce was dominated by engineers who undoubtedly saw the future accurately but were less concerned than they might have been with the financial consequences of the strategic technical imperatives.

Although Rolls-Royce plc is currently in a healthier state than it was in the late 1960s and early 1970s, with seven engine types in production, it still displays a somewhat erratic underlying financial performance (Table 6.6), similar to that of the model with innovations spaced at three-year intervals in Figure 6.4, not least because of the varying impact of R and D costs, most recently of the Trent engine, a project started around 1987 which entered service in 1994.

Although no longer in receipt of launch aid, Rolls-Royce remains dependent on external financing of its R and D, and thus cannot be considered as fully mature in terms of the model. In his statement in the 1990 annual report, the Chairman of Rolls Royce reported that the gross expenditure on R and

D in the last five years had exceeded £1,700 million, of which £866 million had been charged to revenue account after allowing for contributions from governments and overseas partners, a figure comparable with the £853 million profit over the same period. In 1992, the gross figure was £482 million and the net £229 million, a pattern which the Chairman did not expect to change in the immediate future.

Table 6.6 Rolls-Royce plc: financial performance, 1989–94

	1989 £M	1990 £M	1991 £M	1992 £M	1993 £M	1994 £M
Turnover	2962	3670	3515	3562	3518	3163
Trading profit	383	468	335	325	329	309
R and D (net)	(161)	(237)	(216)	(229)	(253)	(218)
Profit on ordinary activities before tax	233	176	51	(184)	76	101
Profit attributable to shareholders	192	134	24	(202)	63	81

Source: Annual reports

Continuous improvement, experience curves, and options

Apart from the problem of being too small, the examples of British Aerospace and Rolls-Royce bring out the importance of the concepts of continuous improvement and the value of options.

The 'education' element of launching costs, as defined by British Aerospace, specifically recognises that as experience builds up in making a product, then its manufacture becomes more efficient. With aircraft, which are built in small numbers, largely individually, the effect is large, but it has also been well documented in mass-produced products such as cars, integrated circuits, and broiler chickens. The graph of unit cost against numbers made is often known as the 'experience curve', a concept particularly associated with the Boston Consulting Group, which found (Boston Consulting Group 1975) that each time the accumulated experience of manufacturing a particular product doubles, the total unit cost can be made to decline by a characteristic factor, normally in the region of 20 to 30 per cent.

The experience curve concept also links to one of the major precepts of the more recent total quality management (TQM) movement, that of 'continuous improvement'. This is essentially a generalisation to all aspects of company operation of the concept of incremental product innovation, in which the cumulative effect of many small improvements can be very significant, and a major source of competitive advantage. From the financial point of view, the great attraction of improvements of this kind is that they usually cost very little. The learning curve is a contributor to the returns from an innovation.

The Rolls-Royce case, where the missing generation of engines contributed to the eventually excessive costs of developing the RB211, illustrates the value of options. If the RB178 programme had been continued, even as a demonstrator from which no immediate return would have been gained, it would probably have reduced the problems of the RB211 considerably, with possibly major influence on the history of Rolls-Royce. In the case of major chemical plants, Robert Malpas, Chairman of Cookson and an engineer with board-level experience in ICI and BP, has drawn attention to the need to think of 'the plant after next' and how experience with a first plant of a new type, although it may be unprofitable, lays the groundwork for successor plants which will be profitable (Malpas 1991).

The importance of the options concept is also illustrated by the consequences of the introduction by the firm of Pilkington of the float glass process, briefly mentioned in Chapter 1. The development of this process, which completely revolutionised the production of flat glass, and has now displaced all but the smallest production plants of other types throughout the world, has been well documented (Pilkington 1969). In the float process, molten glass is spread on the surface of a bath of liquid tin where it cools to produce sheets with surfaces which are parallel and both flat (at long range) and smooth (at short range).

The story of float glass is most familiar as one of persistence in the face of technical obstacles and, to a lesser extent, as one of successful exploitation through licensing. But it also illustrates the opening up of new possibilities in addition to those at which it was originally aimed. The process replaced two previous processes, drawing and grinding. In the first, sheets of glass were drawn vertically from the melt. The surfaces, being free to smooth themselves under surface tension forces, were smooth at short range, but the long range flatness of the sheet could not be controlled precisely, and the product was suitable only for domestic window glass and other optically relatively undemanding applications. For more demanding applications, such as shop fronts and large windows, plate glass was required. The plate glass process, which in its most developed form ground both surfaces of a sheet of raw glass simultaneously, was capable of producing glass whose faces were accurately flat and parallel over large distances, but whose local surface smoothness was inferior to that of sheet glass.

The float process was originally conceived as a replacement for the production of the relatively expensive plate glass. Substitution was helped by a fortunate accident of nature, which resulted in the equilibrium thickness of the glass layer being very close to that of the most popular thickness of plate glass. The technology was, however, soon developed in order to produce thinner glass and to lower the production costs to the point at which the float process also displaced the drawing process (Ray 1984).

The next stage of development was to take advantage of the very special production conditions needed for float glass, in which the lower surface is in intimate contact at a high temperature with liquid tin, and the upper surface is in a controlled atmosphere which has to be maintained to prevent oxidation of the tin. Further, the glass is held for several minutes at a temperature at which it has appreciable electrical conductivity, so that it was possible to develop the 'electrofloat' process, in which an electrical circuit is established between the lower surface of the glass ribbon and a stable layer of molten metal placed on the horizontal top surface. Passage of a current forces metallic ions from the layer on the top surface into the glass, and subsequent exposure to the reducing atmosphere converts the ions to colloidal metal particles. These particles produce optical effects which depend on their nature and size (Robinson 1970). The electrofloat process has since been exploited to manufacture a range of tinted glasses for solar control, decorative effects, and opto-electronic applications, and developments continue.

This range of options opened up by the float process is an example of the so-called 'reverse product cycle' in which first costs are lowered, then new features are added, and finally new products are launched – in the exact reverse of the normal product cycle described above in connection with the automobile industry. The reverse product cycle was originally thought to be associated with the introduction of new technology into services, but it seems to be characteristic of the application of new process technologies.

USE OF THE MODEL FOR PERFORMANCE MANAGEMENT

In principle, the cash flow curve for an innovation provides the answer to two of the most frequently-asked and difficult questions in innovation management: how much should a company spend on innovation?; and what does the company get back for what it spends? The questions are closely related, with the first expressing the point of view of managers looking inward, and the second that of those looking outward.

The Hewlett Packard example quoted in Chapter 4 is an example of using the curve as an aid to managing individual projects, while the work of Grabowski (1991) and others in the pharmaceutical industry is aimed more at the strategic implications for the industry as a whole.

It is also possible to use the financial model of a company as a series of innovations characterised by the cash flow curve for strategic management in an individual company. The basic information needed is the shape and size of the curve for the company's average innovation. This information can be built up by measuring the expenditure in any one financial year and the return being obtained in the same year from previous innovation expenditure. The ratio of return to expenditure gives some measure of efficiency of the innovation process as practised by the company. Given the company's strategic target in terms of growth, the inverse ratio, named above as the innovation ratio, provides a guide as to how much to spend in the future.

This type of approach is likely to be of most value to companies which are of moderate innovation intensity. In aerospace and pharmaceuticals, innovation is the heart of the business which revolves around it; in retailing, although the number of new products is very large – for example, Sainsbury's annual report for 1993–94 records that it introduced about 1,000 new products into its supermarket chain during the year, out of a total range of about 18,000 – the expenditure on them is not large in relation to the business and is, in any case soon recouped or written off as abortive. In the materials, general engineering, and many other industries, however, the decision on how much to spend is more difficult, and the issue of value for money is more often raised.

The Pilkington example, described in Chapter 4, is a response to the second question, in the context of communication with major shareholders and analysts. As we have seen in Chapter 2, investors and others have access to R and D expenditure figures, and so the Pilkington methodology has been developed to measure the return made possible by R and D, rather than the efficiency of the innovation process.

A growing number of companies are measuring and reporting on the percentage of sales income from new products. The rate of turnover of products varies in different industries, but the most popular approach is to report the proportion of sales from products introduced in the last five, or sometimes three, years.

The efficiency of the innovation process

It has been recognised for many years that the return on innovation expenditure, and particularly on R and D expenditure, is not uniform over time for an individual design configuration. In the early stages of a new development rapid progress is possible. The rate tends to slow down, and eventually to become so low that further effort to improve the product or process is not worth while. This 'S-curve' effect, which is essentially part of the fine detail within the broad model presented above, based on the major innovations only, is of general application, and it is important that managements recognise it sufficiently far in advance to start a new generation in time for it to be introduced before the old one reaches the end of its life.

Other things being equal, the bigger the return obtained from innovation expenditure, and therefore the lower the innovation ratio, the better for the company. Provided that the flow of new products can be maintained at a satisfactory rate, there is no point in spending more on innovation. The model provides a way in which managements can distinguish, for their own use and perhaps for communicating to shareholders, between a reduction in expenditure arising as a result of increased innovation efficiency and a similar reduction arising from a desire to cut costs, regardless of the future.

The relationship between technological and financial classes of innovation

The conventional division of innovation into fundamental, radical, and incremental categories is based on technological characteristics. One way of expressing the distinction is to say that fundamental innovation requires that the relevant textbooks be completely rewritten, radical innovation requires a new chapter, and incremental innovation necessitates only minor modifications to existing chapters.

From the financial point of view, a more relevant distinction would be that fundamental innovation requires that the company introducing it seeks major financing from outside, as in aerospace; radical innovation makes demands on the company which depress its profits for a period, as with Pilkington's development of float glass and Glaxo's development of both Zantac and its marketing; and incremental innovation can be taken in its financial stride.

Although the financial and technical definitions may agree in many cases, they may not always do so. On the financial definition, fundamental innovation is a characteristic of new companies, with radical and incremental innovation being the province of existing ones. Schumpeter (1911) envisaged innovation as being introduced by new companies, and it certainly seems to be true that such companies are predominantly involved in fundamental technological innovation which existing companies often overlook. IBM's move from mechanical to electronic office machinery is, however, a counter-example.

Since the exploitation of fundamental technological innovation possibilities often requires very little capital in the early stages, the explanation of this phenomenon is more likely to be the larger companies' lack of the appropriate intellectual capital, such as knowledge and adaptability, to take advantage of the opportunities. If these companies are to survive into the indefinite future, they must incorporate the necessity for innovation more comprehensively into their corporate consciousness.

Questions for discussion

1 If Rolls-Royce had known at the beginning what the development of the RB211 was going to cost, what should it have done?
2 The warrants for shares in British Biotech have an exercise price of £5.25, which was about the current price when its 1995 accounts were published. There were forty-eight million shares in issue, valuing the company at around £250 million. In the light of the characteristics of its industry, are the shares accurately valued, and when are shareholders likely to receive a dividend?
3 How does the development of Airbus (Figure 6.6) fit with the model of a company as a series of innovations?

Chapter 7

Innovation in the national policy context

The emphasis in this book up to now has been on innovation from the company perspective. In this final chapter we look at innovation from the national policy viewpoint, especially that of governments. The topics covered are surveys of R and D and innovation activity, the economics of innovation, government support schemes, and current developments in accounting and reporting which are of relevance to innovation.

STATISTICS ON RESEARCH AND DEVELOPMENT EXPENDITURE

Quantitative information about innovation has in the past been almost entirely confined to its R and D components. These components have been reasonably well documented at the industrial sector level in the major industrial countries for several decades, and the information about government expenditure is equally comprehensive. More wide-ranging surveys of innovative activity have begun to appear since about 1990, but are still fairly embryonic, as their authors wrestle with the problems of measuring such a broad concept.

Information at the individual company level, confined to the R and D component as a single figure for total expenditure, has been available in the US for many years, but in the UK the coverage has become reasonably comprehensive only since disclosure of R and D expenditure became a requirement of the relevant accounting standard (SSAP 13 1989) for accounts for periods beginning on or after 1 January 1989.

Broad figures for UK government expenditure on R and D are published on an annual basis, and reasonably detailed accounts, which are prepared on a department-by-department basis, have been collected together in an Annual Review since 1983.

The OECD, which developed the standard definitions of research and development, regularly collects and publishes data from its members. These publications are the major source of international comparisons, although the EC has recently entered the field (European Commission 1994a).

Apart from the expenditure figures published by individual companies, the industrial data are based on surveys, whose questionnaires inevitably take time to prepare, distribute, complete, collect, and analyse. Despite their limitations in terms of coverage and topicality, it is these surveys which form the foundation for academic and other studies on which policy recommendations and actions are based. Their influences are therefore considerable, and we examine the major sources of data below.

Surveys of industrial R and D expenditure

The first published survey of industrial expenditure in the UK, relating to 1945–46, was carried out and published in 1947 by the FBI, an association of manufacturers which was the largest constituent of the CBI when it was formed in 1965. This survey found 420 firms spending a minimum of £1,000, making a total of £21.8 million and, on the basis of an estimated coverage of 75 per cent, estimated that the total for UK industry as a whole was £30 million (about £600 million at 1995 prices). The report refers to a previous, apparently unpublished. survey which had found 422 firms spending £1.7 million in 1930, 484 firms spending £2.7 million in 1935, and 566 firms spending £5.4 million in 1938. Although prices approximately doubled between 1930 and 1946, there was clearly a substantial increase in activity over the period.

This increase continued until about the mid-1960s, by which time the government had taken over the collection of industrial R and D expenditure statistics. Since that time, the industrial employment on R and D, which is a more straightforward measure of effort than is expenditure in a time of rapid inflation, has remained approximately constant. Figure 7.1 shows that the number of support staff in industrial R and D declined significantly over the period 1968–93, as did the number of technicians. The number of professionals (qualified scientists and engineers), however, increased. The period has been one of considerable expansion in higher education, and the change in the relative numbers of technicians and professionals probably owes more to this cause than to changes in demand for particular skills. This explanation is supported by the fact that the total of technicians, engineers, and scientists has remained approximately constant, with some variation arising from economic cycles.

As Figure 7.1 makes clear, the surveys were performed annually in the late 1960s. Government enthusiasm for the surveys has varied over the years, and in the 1970s a policy of carrying them out at three-year intervals was adopted. In early 1981 the CBI became concerned about the lack of information on the effects of the current recession on R and D, as the last survey had been in 1978 and the results of that planned for 1981 would not be available until 1983. The CBI therefore carried out, in consultation with the government, a sample survey of sixty-four firms, exploiting the fact that the distribution of industrial R and D expenditure is extremely skewed. It was estimated that the responses from the CBI sample survey covered about 40 per cent of the relevant expenditure.

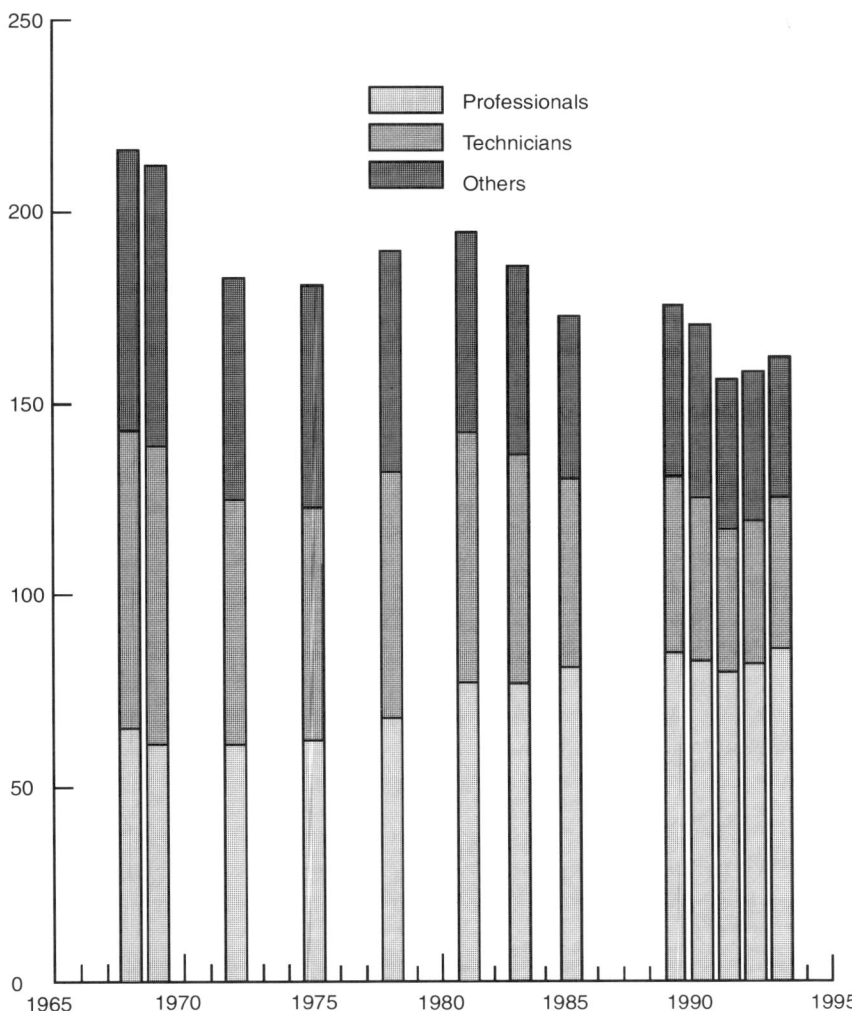

Figure 7.1 UK industrial R and D employment, 1968–93
Source: BSO (1988), CSO (1991, 1995)

Encouraged by this experiment, the government adopted a policy of carrying out major surveys at four-year intervals, interspersing them with annual
sample surveys. Full surveys were carried out for expenditure in 1985, 1989,
and 1993, after which the policy again changed, with the full surveys being
abandoned in favour of annual sample surveys.

A potential user of the statistics will find them scattered over a range of publications, with occasional compilations. The results of the full surveys from 1975 to 1993 were published by the BSO and CSO as the MO14 series of Business Monitors. They give a good deal more information about the composition of industrial R and D expenditure and employment than is available elsewhere. There are breakdowns of expenditure by type (capital, salaries and wages, and other current), source, and region; and employment by level of qualification. The figures are also given in current cash terms and corrected for inflation. The results of sample surveys have most recently appeared as CSO Bulletins.

Interpretation of expenditure data is hampered by the inflation adjustments mentioned above, gaps in coverage, uncertainties about definitions of R and D, changes in the definition of sectors, and the movement of companies between the public and private sectors through nationalisation or privatisation. This last effect has been allowed for in the presentations, but the other effects mean that the trends in the figures are more reliable than the absolute values.

The 1975 survey included companies with more than 100 employees, but this figure was raised to 200 for the 1978 and subsequent surveys. Although the effect on total expenditure is probably very small, any changes in a sector which the government regards as being important to the economy are not being documented. There is concern that this gap results in the software sector, in particular, being inadequately covered. Another change between 1975 and 1978 was to move from counting employees who spent more than half the week on R and D to counting full-time equivalents (FTE), to accord with the OECD definition.

The R and D Scoreboard

Once SSAP 13 (1989) required disclosure of expenditure on R and D by larger companies, it became possible to collect and analyse the figures on a regular basis. The DTI, through its Innovation Unit, commissioned a commercial organisation specialising in collection and analysis of data from company reports to provide this service, with the result that an 'R and D Scoreboard' has been published since 1991. As with other reports, the Scoreboard has evolved to some extent. Its basis is the R and D expenditures reported by UK companies. Tables are given (DTI 1995) of the expenditure in size order both in total and by sector, with figures for the percentage change since the previous year, expenditure for the previous three years, for the proportion that the current expenditure represents of sales, profit, and cost of funds (interest plus dividends), and for the company's P/E ratio.

The Scoreboard also includes data on major foreign companies. Some commentators have pointed out that UK companies seem to spend less on R and D than their overseas competitors, both absolutely and relative to other measures, particularly to profits and dividends. This comparison ignored the differences, discussed above, between countries both in accounting practices and financing structures. The 'cost of funds' heading, newly-introduced for 1995, allows for the second of these differences, but accounting differences remain.

The Scoreboard is a useful compilation, but its implication that spending money on R and D (or investing in R and D, in the terminology that the Innovation Unit prefers) is unambiguously worthwhile is an unfortunate one, reflecting the linear model of innovation which has been officially discarded but whose influence remains.

The highly-skewed distribution of industrial R and D expenditure, mentioned above as the foundation for sample surveys, shows up very clearly in the Scoreboard. Glaxo alone, as the heaviest spender, accounted for 11.5 per cent of the total expenditure in the 1995 Scoreboard, with the next four companies spending 29 per cent between them, and the following five another 19 per cent. The top ten spenders thus had a 59 per cent share of the total.

The figures produced by the Scoreboard are based on company accounts, which report expenditure borne by the company itself, whereas the government surveys are based on R and D performed, regardless of the funder. For most companies, there is no appreciable difference between the two figures, but for those in receipt of government support, notably in the defence sector, the difference may be marked. The example of Rolls-Royce has been quoted in Chapter 6, and another example is GEC, which also pays for less than half its R and D expenditure from its own funds.

The Director survey

In 1994, the *Director*, the magazine of the Institute of Directors, published a table of company R and D expenditure as a proportion of value added, thus filling a gap in the available statistics. The survey was repeated in 1995. The figures were based on aggregate results for four years, rather than one, and also included a figure for R and D expenditure as a percentage of sales.

As with the Scoreboard, there was an unfortunate implication in the commentary on the table in both years that devoting a high proportion of value added to R and D was in some way commendable. Not surprisingly, in the light of the financial model presented and discussed in Chapter 6, the companies which came out at the top of the *Director* list were small R and D-intensive companies in the early stages of growth. They were followed by pharmaceutical companies and then by instrument companies.

The company which came top of the *Director* list in both 1994 and 1995, Tadpole Technology, a maker of advanced portable computers, had spent 59.5 per cent of its added value on R and D in the four years prior to the 1994 survey, and 15.4 per cent of its sales. By 1995, both figures were down, to 41.5 per cent and 9.8 per cent respectively. Value added as a proportion of sales had dropped from 26 per cent to 24 per cent. It was not entirely surprising, therefore, that Tadpole Technology subsequently reported an unexpectedly high loss, with consequent severe effects on its share price.

Government and national R and D expenditure

The Annual Review of government R and D expenditure, which started as a modest fifty-two page publication in 1983 and grew to 300 A4 pages by 1993, was replaced in 1994 by the annual 'Forward Look' of government-funded science, engineering, and technology. This publication (OST 1995b) consists of three volumes, respectively concerned with an overview, forward look statements from Research Councils and government departments, and statistics. The Research Councils and departmental volume reports on outturn for the previous two years, the estimated outturn for the current year, and the plan for the following year.

Statistics for government expenditure include a category of 'strategic' applied research, defined as applied research where the work, although directed towards practical aims, has not yet reached the stage where eventual applications can be clearly specified. The term 'strategic research' was introduced into UK science policy in 1971, at a time when there was considerable debate about definition and purposes of research, resulting in many definitions being put forward (Rothschild 1972).

The statistics volume of OST (1995b) also includes the main results from surveys of industrial R and D, and is the most comprehensive source of information available.

Interpretation of the figures

The figures have proved most controversial when used in aggregated national form to calculate the ratio of R and D expenditure to GDP, which is the figure most used for international comparisons; and the division between civil and defence expenditure, which is a matter of dispute between those who think that the government should support civil R and D and those who think it should not.

As experience has built up, confidence in the accuracy of the figures has gone down. For example, in 1973, the total UK R and D expenditure for 1964/65 was reported to have been 2.653 per cent of the GDP. In 1985, the figure for 1964 was reported to have been 2.32 per cent. The discrepancy seems to be accounted for by changes both in the estimate of the GDP (which is subject to change in the light of new information) and the method by which it is calculated (of which there are three – based respectively on income, expenditure, and value added – which should agree but do not). In view of these problems, more recent publications have been content to quote two significant figures only. The UK ratio has remained at around the 2.3 per cent level which it reached in the mid-1960s, although its make-up has changed significantly over the period.

The definition of R and D or, more accurately, the interpretation of the OECD's 'Frascati' definitions (OECD 1994) has been another source of dispute. It was maintained by some commentators that much of what the Ministry of Defence classified as R and D was not the sort of activity which others would recognise as such. It also emerged from the Annual Review that the defence industry was reporting that it had received only about half as much R and D support from the government as the government expenditure figures showed it to have provided. The House of Lords Select Committee on Science and Technology investigated these problems, and published two reports (HL 1990, HL 1992). A critical review of the available figures has been given by Walshe (1992).

The Lords Committee found that much of what the MoD classified as R and D was indeed not true R and D, and recommended various improvements in the collection and dissemination of MoD statistics. It also pointed out that international comparisons of R and D expenditure were distorted by the inclusion of, probably exaggerated, figures for defence R and D, and recommended that the practice of quoting a single figure for national R and D expenditure should be abandoned in favour of separate figures for defence and civil expenditure.

Probably the most significant finding of the Lords Select Committee was that most performers and funders of R and D did not use Frascati definitions for their own purposes, and had some difficulty in translating their own internal data into Frascati form for completing survey questionnaires. The Committee pointed out that definitions were only a means to an end and need not be the same for all purposes, as long as they were compatible by being related to the Frascati versions. The Committee commented that for comparison purposes, the fewer categories the better, and recommended that for international comparisons, only the two categories, research and development, should be used, with no subdivisions of research. For national policy, the distinction between applied and basic research was useful. The Committee opposed the use of 'strategic' research as a separate category, arguing instead that its inclusion in one of the other categories should be studied by OECD.

OECD responded to the development of interest in a category of strategic research by splitting its basic research category into pure-basic, and oriented-basic (OECD 1994). UK practice has been to regard strategic research as a sub-division of applied research. This difference illustrates the wisdom of the Lords Committee's comment that definitions depend on the purpose for which they are made. The government regards strategic research as part of applied research, in accordance with its definition of applied research as research which has been funded because the funder wants the answer. The OECD seems to be inclined to the view of the performer of the research, who regards research as basic if it can be published in learned journals.

Defining R and D, and particularly sub-dividing it into categories, has some of the same difficulties as defining profit. These difficulties arise from imperfect recognition that the definition depends on the point of view of the user and the purpose for which it is to be used. Since there is no point in performing R and D, apart from basic research about whose definition there is little dispute, unless some return is obtained from it, it has gradually been realised that rather than pursue definitions down some increasingly esoteric path, it is more more fruitful to develop techniques for measuring those returns, or in other words, for measuring innovation.

INNOVATION SURVEYS

Although the OECD recognised from the early days that R and D was only part of the innovation process, there has been very little systematic work on collecting information about the expenditure on those other parts, and still less on the returns from the total innovation expenditure, until the last few years. The main sources of information about industrial innovative output were the patent offices of the major industrial countries, and especially the US Patent Office; while the main sources of information about the outputs of research were the papers published in learned journals and analyses of the citations which those publications attracted. Both these sources have severe defects as measures of innovative output: the importance of patents is very different in different industries, and the publication-based measures are very indirect indicators of innovative activity. Neither has any direct link with financial success.

The main reasons why there was a long delay before interest began to build up in measuring innovation seem to have been the intrinsic difficulty of deciding what to measure and how to measure it, and a belief that, in any case, measurements of R and D expenditure captured its most important element. This latter belief reflected the linear model of innovation, discussed in Chapter 1 which, although long abandoned in industry (Budworth 1973), persisted in academic and government circles until around the late 1980s, and has still not lost its influence on thinking.

The CBI Innovation Trends Survey

In the UK, the CBI was again a pioneer in seeking to extend the scope of measurement, starting its Innovation Trends Survey in 1990. The sixth in the annual series was published in 1995 (CBI 1995). As its name suggests, the survey does not collect detailed figures but concentrates mainly on changes. One figure for which respondents are asked, however, is the percentage of turnover devoted to innovation. This figure rose in the 1995 survey (of activity in 1994) to 6.7 per cent for manufacturing, from an average of just over 4 per cent in 1991, 1992, and 1993. (It will be noted that the latest figure is close to that used in the financial model of Figures 4.2, 5.1, and 6.1.) For non-manu-facturing, there was an even greater rise in innovation expenditure, from around 6 per cent to around 11 per cent. Other marked changes in 1994 were an increase in collaboration, a rise in the effect of regulations as both incentives to and constraints on innovation, and a shortening of development times. The causes and significance of this last change were discussed in Chapter 6.

The Central Statistical Office review

The first official UK publication on innovation indicators was a review by the Central Statistical Office of those for activity both in industry and in science and technology (Doudeyens and Hayman 1993). This survey gave figures for the UK and its major industrial competitors for various time periods of:

(a) Enrolments per field of study at first degree level.
(b) Total R and D labour force.
(c) Gross expenditure on R and D as a percentage of GDP.
(d) Structural indicator of business R and D, defined as the weighted average of the R and D intensity of each manufacturing industry, expressed as a proportion of the international average for that industry.
(e) Distribution of R and D effort and patents by sector of industry.
(f) Grants of European patents by country of origin.
(g) Percentage of US patents.
(h) Papers published and citations.
(i) Technology balance of payments (defined by payments for intellectual property rights).
(j) Stock of installed industrial robots.
(k) Percentage of exports by intensity of R and D.

The OECD Guidelines

Other countries had made rather more progress than the UK, and the OECD was able by 1992 to produce guidelines for innovation surveys (OECD 1992). These guidelines cover:

(a) Objectives of innovation.
(b) Factors assisting or hampering innovation.
(c) Identifying innovating enterprises and numbers of innovations.
(d) Qualitative aspects of innovation (novelty and nature).
(e) Impact of innovation on the performance of the enterprise.
(f) Diffusion of innovation.
(g) Patents and technology balance of payments.
(h) Measuring the cost of innovation.
(i) Suggested breakdowns.

The Community Innovation Survey

It can be seen that the field covered by OECD (1992) is wide, reflecting the managerial and strategic factors affecting innovation decisions, as well as the purely financial aspects. OECD (1992) includes a suggested list of questions for a survey which has been adopted by the European Commission as the basis for a Community Innovation Survey (CIS) (European Commission 1994b). The CIS includes questions on R and D expenditure and its split between products and processes; on total innovation expenditure and its split between R and D, patents and licences, product design, trial production and related activities; market analysis; and others. Other questions cover the distribution of sales between products in the introductory, growth, maturity, and decline stages of their life cycle; and the distribution of sales amongst products which were unchanged, incrementally changed, or significantly changed during the previous two years.

The prospects are thus that a considerably greater and potentially more useful body of information about the financial and other aspects of innovation will be available in the future.

THE ECONOMICS OF INNOVATION

Innovation and technical change

There is a large literature on the economics of innovation, most of which is outside the scope of this book. The purpose of this short section is to explain some of the leading ideas in the field, and particularly how they affect the financing of innovation in a company. Three recent compilations survey the field. Dosi *et al.* (1988) has twenty-eight contributions from the editors and others which were developed collaboratively and therefore dovetail into a critique of current economic theory and the inadequate attention given by it to technical change. Freeman (1990) is a compilation of classic papers in the field with an introduction by the editor. Stoneman (1995) is also a multi-author work, this time consisting of thirteen chapters, including one on finance and technological change which takes a more severely theoretical and mathematical approach than this book and concentrates mainly on the problem of making the investment decision for a major long-term R and D project.

Schumpeter's classification of invention, innovation, and diffusion (Schumpeter 1911) remains as one of the foundations of the economics of innovation. Economic studies are still made on these three aspects, and the influence of the tripartite division can be seen in the format of the OECD guidelines described above.

For many years, Schumpeter remained something of a lone voice, outside the mainstream of economics. But interest in technical change began to increase when it was discovered in the late 1950s, by careful empirical study of the data, that most of economic growth could not be accounted for by inputs of the three factors of production of classical economic theory – land, labour, and capital – and must therefore come from some other source, which was dubbed 'technical change', whose effect was to increase the efficiency with which the inputs were converted to outputs. The processes by which this improvement came about and their economics thus became of interest.

Economic studies of innovation

One of the problems which these studies threw up was that classical economic theory assumes that economic actors – firms – are rational, pursuing profit in the most efficient way. The uncertainties of innovation, and the difficulty of the originator in appropriating the fruits of the innovation, made rational choice more difficult, and possibly resulted in innovative activity being at a level which was optimal for the individual firm, but sub-optimal for the economy as a whole. Economists therefore became interested in why some firms were more interested in innovation than others, what the incentives to innovate were, and whether government subsidy for some aspects of innovation might be be beneficial overall.

A second problem for the classical theory was that it is concerned with predicting the equilibrium state of an economic system. Once innovation is introduced as a major factor, however, the system becomes dynamic and evolutionary. Dosi *et al.* (1988) summarise the work of economists who have developed dynamic theories. Classical theory had reached a very high degree of mathematical sophistication, based on solving complex systems of simultaneous equations describing the equilibrium relations between economic variables, and it is only since the development of computer modelling approaches that techniques have existed to deal with the differential equations which are needed to represent a dynamic system. Much work remains to be done on using these techniques.

One branch of the economic tradition is based on analysing published statistics in a search for correlations, and is thus heavily dependent on the availability of statistics. Since, as we have seen, those on innovation are sparse, economists of this tradition have been forced to use R and D expenditure as a proxy for innovative activity and patents as a proxy for innovation output, with consequently uncertain results. Other problems with this approach are the difficulty of determining the direction of causality, and the limitations of the accuracy of the data.

A second economic tradition still operates at the macroscale of sectors, but concerns itself with the realities of technology and the way in which it is dealt with by firms. This approach is not afraid to depart from idealised assumptions of perfect rationality, complete information, and maximisation of profit, and accepts that equilibrium may be a long time coming. This school has produced some work which is illuminating to those not intimately concerned with the sectors which have been studied.

The third major tradition is the empirical, case-study-based approach, whose focus is the individual firm. This tradition has provided some very valuable insights into the rather accidental factors which affect innovation, and has corrected some of the simplified assumptions in the models used by other schools.

The general picture which emerges from many studies is of a system in which the history, management, and aspirations of the individual company count for more than the actions of government. Nonetheless, the structure of the national economic system, consisting of government, companies, financial institutions, and the legal framework does have an effect on its performance. Studies on national systems for financing innovation have thus become of interest (Nelson 1993, OECD 1995).

Innovation and industrial strength

A major product of the statistical analysis school of economic research was published under the above title by Cox and Kriegbaum (1989). Rather than study existing data in the form in which they were collected, their aim was to use them as the basis for an analysis of the way in which industries divide their expenditure between current products and future products, and to examine the influence on this division of the environment in which the industries operate, and the economic results which then follow. Total innovative effort was thus put into the context of the financial structure, essentially in the form of a value added statement, of the individual sector in which it occurred, and was interpreted in the light of the economic environment and consequences.

Cox and Kriegbaum (1989) give cost breakdowns for the mechanical engineering, motor vehicles, chemicals, textiles, and electronics industries in the UK, Germany, Japan, and the USA, over the period 1972–1985. This period included two major oil price shocks and various attempts by governments to control wages and prices. Confirming an earlier, more limited, study, they found that the pattern of costs in each industry was stable both over time and between countries. It was therefore possible, in addition to the detailed annual figures for each country, to give a normalised breakdown for each sector. These breakdowns are shown in Table 7.1.

The main adjustments which Cox and Kriegbaum (1989) made to the raw data were to calculate the innovation expenditure, using the observed R and D expenditure and such information as was available about the ratio between the two, and to adjust the raw employment costs for the innovation component (cf Table 2.12). As the arithmetic of Table 7.1 shows, the 'disposable funds' were what remained after costs for current output had been met. To these were added the 'other funds', which were extra contributions other than from sales of goods produced, to make a total of funds at the disposal of management.

('Other funds' arose from profits for merchanted goods, sales of surplus equipment, income from investments, investments from abroad, and government grants.) Of these disposable funds, some were devoted to investment, in both intangible (R and D and innovation) and tangible (fixed capital) forms, leaving a surplus or deficit. The value added figure is also given, calculated as sales less purchases for current output.

The authors concluded from their study that West Germany (as it then was) had pursued policies which, on the whole, come nearest to the optimum. They pointed to the changes of policy and lack of consistent government support for industry which characterised the UK; to an excessive reliance on industrial services in the USA; and to the low margins and excessive dependence on new capital injections of Japanese industry.

The main conclusions, which were particularly directed at the UK, were that more attention should be given to the intellectual assets of companies, and to the revenue expenditure on training, research, and innovation generally, which sustain those assets.

Table 7.1 Cost structures for five industries

	ME*	El*	MV*	Chem*	Tex*
For current output:	86	74	84	85	94
Materials	40	38	50	50	50
Employment costs	32	25	22	20	25
Industrial services	4	3	2	2	4
Other services	6	2	6	8	9
Interest payments	2	2	2	2	2
Taxes	2	4	1	3	4
Disposable funds	14	26	16	15	6
Other funds	4	6	8	5	4
Available funds	18	32	24	20	10
For investment:	10	20	12	14	6
R and D and innovation	6	15	8	9	2
Fixed capital	4	5	4	5	4
Surplus/deficit	8	12	12	6	4
Value added	50	57	42	40	37

Source: Cox and Kriegbaum (1989)
* ME: mechanical engineering; El: electronics; MV: motor vehicles; Chem: chemicals; Tex: textiles

Statistical studies at the firm level

A more recent study at the firm level (Clayton and Carroll 1994) confirmed the conclusion of Cox and Kriegbaum (1989) about the importance of intellectual assets. Clayton and Carroll (1994) used the information about 3000 business units (distinguishable businesses, mainly embodied in larger companies or groups) in the proprietary PIMS database (Buzzell and Gale 1987) and a database of company information used by the European Commission. The

PIMS database covers about 400 variables for each unit, over a minimum of four years, enabling extensive statistical tests to be carried out.

The three major conclusions of this study were:

1 Intangible factors are more powerful determinants of medium-term business growth than the tangible factors measured.
2 Innovation and intellectual property are the strongest drivers of competitiveness. There are clear statistical links – varying by market situation – which link R and D, management capability, intellectual property, innovation, customer preference, and market focus to growing market share, value added, and jobs.
3 Relative costs and productivity affect growth and profits, but labour productivity gains which are mainly achieved by 'capital deepening' – substituting capital for labour – usually destroy both jobs and profits.

Innovation is defined in the PIMS database – in accordance with the Innovation Unit definition used in this book – as percentage of revenue in a business derived from products and services introduced in the last three years. Intangible investment included not only R and D but also expenditure on the sales force, on promotion, and on advertising relative to competitors.

An important result of this study – confirming the views of practitioners of innovation, who were not in the grip of the linear model of innovation – was that the correlation between R and D spend and sales or market share growth was so weak as to be statistically insignificant. The translation of R and D to financial success was shown to depend on intermediate steps, particularly those related to meeting customer demands.

Returns to R and D expenditure

This last result of Clayton and Carroll (1994) throws valuable statistical light on the subject of returns to R and D expenditure, which has been a subject of debate for many years. The justification of R and D expenditure, and especially that by governments, as a contributor to economic growth has naturally led the curious to ask whether these benefits have in practice been achieved. Although attempts have been made to answer this question at both the national and the company level, it is not surprising that no very conclusive answers have been found.

An early UK attempt at the national level was by Byatt and Cohen (1969) who proposed a methodology for tracing the benefits of scientific (by which they meant what would nowadays be called basic) research through the economic system. This pioneering attempt to design what was essentially an innovation survey was not followed up.

A review in 1991 (Smith 1991) concluded that there was evidence that the rate of return to companies (the private rate) was significantly lower than the return to the economy as a whole (the social rate), because of interactions between firms, in which an innovation in one sector produced productivity gains in another. Smith concluded that there might well therefore be under-investment in publicly-funded R and D, but that the lack of data and the problems of measurement hindered progress. One of the problems was that existing data concentrated almost entirely on R and D alone, ignoring other inputs to innovation, while another was that there was little information about inter-sectoral transfers, through which many of the benefits of innovation accrued to the economy. The OECD questionnaire, with whose development Smith was involved at the time of his review, was designed to fill these gaps.

The subject was reviewed again by OST and PREST (1993) who also came to the conclusion that lack of data hampered any sound assessment of quantitative financial returns, although there was some useful work on qualitative assessment of R and D projects.

If it is accepted that R and D is only a part of the total innovation process, and that it is from that total process that a financial return is obtained, then it follows that the question of measuring returns on R and D is not a valid or sensible one to pursue. The returns which R and D has made possible can be ascertained in suitable cases, as has been done by Pilkington (Chapter 5), but it would be difficult to extend that approach to the national R and D effort and the benefits to the economy as a whole, in which the links between R and D and its exploitation are much more complex than in an individual company.

GOVERNMENT SUPPORT FOR INNOVATION

Government financial support for innovation, mainly of its R and D components, has existed in one form or another for very many years. It can take two forms: direct, in which grants are given according to some criteria of national interest; and indirect, through tax concessions.

Direct support

The main features of direct support schemes have been discussed in Chapter 5 where, it will be remembered, there was a warning that the details of such schemes change rapidly, and no written account of them, even from official sources unless hot from the press, can be taken as gospel.

The justification for support schemes is the concern of economists that the amount of R and D performed will be sub-optimal from the national point of view if its financing is left entirely to the private sector. The corresponding danger that government support might lead to a super-optimal expenditure has been less well appreciated, although some cases, such as Concorde, brought it to light. The Concorde example also stresses what economists would call the opportunity cost aspect – resources, particularly in the innovation case those of skilled labour, devoted to one object cannot be devoted to another. With hindsight it can be seen that the effort put into developing supersonic civil aircraft would have been better put into large subsonic vehicles.

The significance of this sort of failure to incorporate R and D into the whole innovation process has gradually become more apparent as consideration has been given to the UK's relative economic performance in the light of the facts that, in the early 1960s, it was second only to the USA in the proportion of the GDP which it spent on R and D, and was far in advance of Germany and Japan, and that even as late as 1981 it was spending more of its GDP on civil R and D than was Japan.

As practice has developed, it has been realised that R and D is not a simple input to innovation, but in many cases is highly specific and therefore best left largely to the decisions of particular managements. Although government has an unquestioned role in financing research of an academic nature – the basic research of the Frascati definitions – it increasingly seeks to influence the areas in which this work is carried out to those areas of strategic or oriented-basic research where potential applications seem to be more likely. It has also sought to promote collaboration, either between industry and academia or between companies, where its role as a broker is at least as important as its role as financial supporter.

Indirect support through the tax system

Governments frequently attempt to influence behaviour through the tax system and, despite evidence that the results are often perverse or even the opposite of what was intended, persist in doing so. Even when they take a resolute stand against fiscal distortion, they are subject to intense pressures to change their minds. R and D expenditure has been a popular object of favourable tax treatment, and several countries have introduced permanent or temporary schemes to encourage an increase in such spending. The UK has, as yet, not done so, but the subject has been raised on numerous occasions and the pressure continues, with advocacy by the Select Committees of both Houses of Parliament and a suggestion that the issue be reviewed once again having been made in the first report of the Technology Foresight Steering Group (OST 1995a).

The present UK position is that current expenditure on R and D is fully allowable as a business expense, and capital expenditure for research is fully allowable in the year of purchase, instead of attracting only the depreciation allowance on all other forms of capital investment. It will be remembered from Chapter 3 that accelerated depreciation allowances benefit cash flow rather than profit. Since the returns on research, in particular, are very long delayed, a concession to aid investment in it is not unreasonable.

The argument for giving favourable tax treatment to R and D is that companies cannot afford to spend as much as they would like to, or should do in the interests of the national economy, given the other pressures on them. Those who think the problem is a permanent one advocate permitting companies to accumulate a credit at some multiple, such as 125 or 150 per cent, of their R and D expenditure: those who think a step-change is needed advocate calculating the credit as any increase in R and D expenditure. The credits are usually set off against tax, but in some schemes are directly refundable, amounting in effect to a subsidy.

Officials of the UK government published a survey of existing schemes in 1987 (Inland Revenue and Treasury 1987). The survey covered ten major OECD countries: Australia, Canada, France, Germany, Japan, the Netherlands, Sweden, Switzerland, the UK, and the USA. The main conclusion, at least from the government's point of view, was that the best available estimates, which were not very good, were that special fiscal incentives increased R and D expenditure by about half the amount of revenue foregone, with the rest going to increase companies' cash flow and post-tax profits.

The report pointed out that the benefit of a tax credit of some kind depends on the tax rate, being worth more to the recipient when the tax rate is higher. Accepting that it would not be effective to attempt to increase the incentive to spend on R and D by increasing the corporation tax rate, the report therefore calculated the pre-tax returns required on an R and D investment in each country, assuming a 5 per cent real interest rate and various R and D 'depreciation' rates, representing the range of times for which the R and D investment would bring in revenue. The results were similar for all countries except Australia, whose 150 per cent allowance for current expenditure and depreciation made R and D investment very attractive, and Canada, whose credits for R and D expenditure were refundable within limits even if no tax liability arose.

The report also found that while some countries were introducing new schemes, others were abandoning or modifying them and that there had been instances of abuse.

ACOST (1993) reported on the issue as one of its last acts before it was wound up. The report recommended a detailed study of the relationship between the UK tax system and the innovative activity of UK-based businesses, giving particular attention to a number of rather technical issues such as definitions, investment in small businesses, and the problems introduced by ACT (Chapter 1), but notably excluding special tax reliefs for R and D expenditure, which it regarded as difficult to target and with effects which were difficult to assess.

A further examination of the issue (Griffith, Sandler, and Van Reenen 1995) identified five difficulties with tax incentive schemes. First, it is difficult to decide at what level they should be pitched and what form they should take, since the benefits, particularly those accruing to the economy in general, are difficult to measure. Second, schemes are subject to change, and consequent lack of confidence in their permanence makes companies reluctant to respond to them, particularly as it is managerially unwise to change the level of R and D expenditure quickly. Third, the definition of R and D is sufficiently imprecise to allow relabelling of other costs to occur if special advantage can be gained thereby. Fourth, schemes, particularly those based on incremental spending, can become very complicated and in practice produce no net effect. Finally, the schemes, particularly if they are tax-based, may fail to help the most deserving cases, such as small struggling companies, because they have no profits to tax and hence no gain from tax reliefs.

Griffith *et al.* (1995) also made the point that tax subsidies can become an internationally-competitive and ultimately self-defeating pursuit. They conclude, somewhat paradoxically, that if policymakers want to increase R and D expenditure, then tax credits are a feasible and attractive option, albeit one fraught with peril.

ACCOUNTING AND REPORTING ISSUES

Accounting, in the sense of preparing the financial statements required by company law and accounting standards, and reporting, through the annual report and accounts, are very closely related, but they are not identical or synonymous. The annual report offers companies an opportunity to explain themselves to their shareholders and other stakeholders, who are increasingly wanting to know more, not just about the past and its financial results, but about the prospects for the future. These prospects, in turn, depend largely on the company's intellectual assets, of which innovative capacity and prospects are a major constituent.

The problems have led to considerable discussion in the accounting profession about the purpose and meaning of company accounts, the very nature of the companies to which they relate, and to various proposals for change. A main theme of this book is that a company is intended to be a continuing entity which must innovate if it is indeed to continue. It must therefore be run and reported on in a way which incorporates innovation as an integral activity.

Making corporate reports valuable

This principle made an appearance in a notably thorough and radical review produced by the Research Committee of the Institute of Chartered Accountants of Scotland (ICAS 1988), one of whose proposals was that company reports should include a statement of relative innovation. This statement would cover the proportion of production that was new and conceived internally and self-generated, and how this performance compared with competitors. It also suggested that the reports should include information about the lead time and effectiveness of R and D, and that this information should be linked to that on relative innovation.

A main principle of ICAS (1988) was that the information required by investors and other stakeholders was essentially the same as that required by management, but in less detail. It therefore proposed that a cash flow statement should be included in the financial statements, and also that estimates of future cash flows should be given.

ICAS (1988) was intended to provoke thought and experiment, rather than as a blueprint for immediate adoption, and was therefore deliberately framed without reference to any existing requirements. Nevertheless, as was discussed in Chapter 2, its proposal for the publication of a cash flow statement was adopted in 1991 by the ASB as its first new standard (FRS 1 1991).

In the more immediate world of the ASB, the response to the demands for change in corporate reporting to accord with the conditions of the modern world takes two, closely-related forms. On the one hand, there has been much discussion about the appropriate treatment of intangible assets. On the other hand, suggestions have been put forward for improved reporting outside the statutory or standards-based framework.

Accounting treatment of intangible assets

The treatment of intangible assets is a difficult area for accountants because it is usually difficult to value them with the degree of certainty, or at least verifiability, which accounts require. The ASB's draft statement of principles includes as one criterion for the recognition of an item in financial statements that it can be measured at a monetary amount with sufficient reliability. Not all intangible assets, and in particular those which the company has generated itself, can satisfy this criterion, and it will be remembered from Chapter 2 that the law forbids the inclusion in company balance sheets of a figure for self-generated goodwill.

Chapter 2 also outlined the problems with accounting for purchased goodwill or intangible assets. Once having arrived at a valuation, there are two accounting choices: to write off the amount against reserves, thus passing the transaction through the balance sheet and not affecting the profit and loss account; or to capitalise the item on the balance sheet and depreciate it to zero over a period by amortisation charges in the profit and loss accounts during that period.

Neither treatment has been found to be satisfactory. Immediate write-off against reserves damages the balance sheets of companies, sometimes to an embarrassing degree. Those who think that the balance sheet should represent the wealth of the company argue that this treatment also gives a false impression since the assets exist and should be shown. Those who think that the accounting treatment should be the same for purchased assets as for self-generated assets of the same kind argue that immediate write-off preserves this equality.

Capitalising and amortising, on the other hand, is supported by those who accept the argument that an asset has been acquired, but also that its value will diminish in time until it becomes a part of the business on the same basis as an equivalent self-generated asset. Others, such as TI (Chapter 2) argue that amortising is not needed because, for example in the case of purchased brands, the value will be maintained by advertising or other promotional expenditure. Since this expenditure, essentially on innovation, is charged to the profit and loss account, to add an amortisation charge would be to double count and therefore distort the picture shown by the accounts.

After a good deal of discussion and consultation, the ASB put forward a solution (ASB 1995) which it hopes will command general acceptance. This is to allow an acquirer to recognise intangible assets separately from goodwill, provided that a fair value for the intangible assets can be measured reliably; to require depreciation of both goodwill and intangible assets which have finite lives, but to allow those with indefinitely long lives to remain undepreciated; and to review the value of goodwill and intangible assets at each year-end. This review would be minimal in extent for assets with lives of twenty years or less, but fuller, including an assessment of future cash flows, for those with a life of over twenty years. Further, ASB (1995) proposes that there should be a rebuttable presumption that goodwill has a finite life not exceeding twenty years.

This approach by the ASB moves further towards adopting the view of the balance sheet as a measure of wealth, while preserving the standard of measurement reliability which is required in statutory financial statements.

The operating and financial review

As mentioned in Chapter 2, the ASB has recognised that the requirements for inclusion in statutory financial statements are such that some important issues cannot be covered in those statements and that users of the statements would be helped by the inclusion of an objective analysis of their contents and how they relate to the underlying business. The ASB therefore formulated a statement (ASB 1993) of voluntary best practice for an Operating and Financial Review (OFR) which is intended as a framework within which the directors of a company can discuss and analyse the performance of the business and the factors which affect it.

The OFR is particularly important from the point of view of innovation which, as we have seen, receives but partial coverage in the normal financial statements, since it is intended to include a discussion, not only of the operating results of the business, but of its dynamics and the investment which it is making for the future. Relevant topics which the ASB envisages as being included within the OFR are the effects on the profit and loss account of new products; of the availability of skilled labour; of patents and other IPR; and of revenue investment in marketing and advertising campaigns, training, R and D, and technical support to customers. It also envisages that there should be a discussion of the 'hidden values' or intellectual capital contained in brands and similar assets which are not included on the balance sheet but contribute to the company's stock market value.

THE WAY AHEAD

Financial accountants, management accountants, and investors are all show-ing considerably enhanced interest in the innovative performance of com-panies and in the value of the intellectual assets to which the capacity to innovate is a substantial contributor. Models of the essentially dynamic process of innovation are available and could be developed further. The OFR offers a structured, but relatively unconstrained, route through which compa-nies can develop their reporting of innovative activity. The way is open for the full integration of innovation into company activity as the competitive condi-tions of the twenty-first century will demand.

Questions for discussion

1 UK industrial expenditure on R and D has remained effectively constant for the last thirty years, as has the percentage of the GDP spent on R and D. What is the explanation?
2 If a tax incentive for R and D expenditure had been introduced in, say 1992, based either on total expenditure or on increases, the biggest beneficiary would have been Glaxo. What are the implications?
3 In the light of the findings of Clayton and Carroll (1994), are there other fiscal incentives which could be used to improve UK innovation perfor-mance?
4 The aerospace industry made a case in 1993 for innovation support from HMG of £100 million per year. Was HMG justified in rejecting it?

Glossary of acronyms

ABC   Activity-based costing
ABPI   Association of the British
   Pharmaceutical Industry
ACOST   Advisory Council on Science
   and Technology
ACT   Advance corporation tax
AIM   Alternative Investment Market
AMT   Advanced manufacturing
   technology
ASB   Accounting Standards Board
ASC   Accounting Standards Committee
ASSC   Accounting Standards Steering
   Committee
BQMA   British Quality of
   Management Awards
BSO   Business Statistics Office
CAD   Computer-aided design
CAM   Computer-aided manufacture
CAPM   Capital asset pricing model
CBI   Confederation of British Industry
CCA   Current cost accounting
CIMA   Chartered Institute of
   Management Accountants
CIS   Community Innovation Survey
CPP   Constant purchasing power
CSO   Central Statistical Office
DCF   Discounted cash flow
DTI   Department of Trade and Industry
EC   European Community, European
   Communities, European Commission
ECU   European Currency Unit
EEIG   European Economic Interest
   Grouping
EIRMA   European Industrial
   Research Management Association
EIS   Enterprise Investment Scheme
EPS   Earnings per share
ESRC   Economic and Social Research
   Council

EVA   Economic value added
FBI   Federation of British Industries
FRED   Financial Reporting Exposure
   Draft
FRS   Financial Reporting Standard
FTE   Full-time equivalent
GAAP   Generally-accepted accounting
   principles
GDP   Gross domestic product
GIE   Groupement d'Intérêt
   Economique
HC   House of Commons
HL   House of Lords
HMG   Her Majesty's Government
HMSO   Her Majesty's Stationery
   Office
ICAEW   Institute of Chartered
   Accountants in England and Wales
ICAS   Institute of Chartered
   Accountants of Scotland
ICFC   Industrial and Commercial
   Finance Corporation
IPR   Intellectual property rights
IRR   Internal rate of return
MoD   Ministry of Defence
MSC   Manpower Services Commission
NCE   New chemical entity
NEDO   National Economic
   Development Office
NHS   National Health Service
NPV   Net present value
OECD   Organisation for Economic
   Co-operation and Development
OFR   Operating and Financial Review
OST   Office of Science and Technology
P/E   Price/earnings ratio
PIMS   Profit Impact of Marketing
   Strategy
PLC   Public limited company

PPRS   Pharmaceutical Price Regulation Scheme
PREST   Programme of Policy Research in Engineering, Science, and Technology (University of Manchester)
PV   Present value
R and D   Research and development
ROACE   Return on average capital employed
ROCE   Return on capital employed
RONA   Return on net assets
RSA   Royal Society for the Encouragement of Arts, Manufactures, and Commerce
SFM   Strategic financial management

SMA   Strategic management accounting
SMART   Small Firms Merit Award for Research and Technology
SPSG   Science Policy Support Group
SPUR   Support for Products under Research
SSAP   Statement of Standard Accounting Practice
SVA   Shareholder value analysis
TCS   Teaching Company Scheme
TDC   Technical Development Capital
TQM   Total quality management
UMIST   University of Manchester Institute of Science and Technology
USM   Unlisted Securities Market
VCT   Venture Capital Trust

References

ABPI (1993) *Pharma Facts and Figures*, London, ABPI.

ACOST (1993) *Innovation and the Tax System*, London, HMSO.

ASB (1991a) *Statement of Aims*, London, ASB.

ASB (1991b) *Statement of Principles, Chapters 1 and 2*, London, ASB.

ASB (1993) *Operating and Financial Review*, London, ASB.

ASB (1995) *Goodwill and Intangible Assets*, Working Paper for discussion at public hearing, London, ASB.

ASSC (1975) *The Corporate Report*, London, ASSC.

Bank of England (1994) *Finance for Small Firms*, London, Bank of England.

Bank of England (1995) *Finance for Small Firms: A Second Report*, London, Bank of England.

Berry, Aidan (1993) *Financial Accounting*, London, Chapman and Hall.

Blaker, G J (1994) *Glaxo's UK Operations over the Next Decade*, London, Foundation for Manufacturing and Industry.

Boston Consulting Group (1975) *Strategy Alternatives for the British Motor Cycle Industry*, House of Commons Paper 532, London, HMSO.

BQMA (1993, 1994) *The British Quality of Management Awards* (sponsored by MORI and Sundridge Park), Bromley, Sundridge Park.

Brealey, Richard A, and Myers, Stewart C (1991) *Principles of Corporate Finance*, 4th edition, New York and London, McGraw Hill.

Bromwich, Michael, and Bhimani, ALnoor (1994) *Management Accounting: Pathways to Progress*, London, CIMA.

Bruce, Robert (1986) *Winners*, London, Sidgwick and Jackson.

BSO (1988) *Industrial Research and Development and Employment*, MO14 1985, London, HMSO.

Budworth, D W (1973) 'What Future for Innovation?', *Technology and Society*, 8, 4–6.

Budworth, D W (1987) *Rewinding the Mainspring*, London, Technical Change Centre.

Budworth, D W (1988) 'Research and Development' in Skerratt, L C L, and Tonkin, D J (eds) *Financial Reporting 1987-88*, 125–138, London, ICAEW.

Buzzell, Robert D, and Gale, Bradley T (1987) *The PIMS Principles*, New York, The Free Press.

Byatt, I C R, and Cohen, A V (1969) *An Attempt to Quantify the Economic Benefits of Scientific Research*, Science Policy Studies No.4, London, HMSO.

CBI (1995) *CBI/NatWest Innovation Trends Survey, Issue 6*, London, CBI.

Clayton, Tony, and Carroll, Charles (1994) *Building Business for Europe*, Final Report to the European Commission by PIMS Associates Ltd and the Irish Management Institute, London, PIMS Associates Ltd.

Collison, David, Grinyer, John, and Russell, Alex (1993) *Management's Economic*

Decisions and Financial Reporting, London, ICAEW.

Coopers and Lybrand Associates (1985) *A Challenge to Complacency: Changing Attitudes to Training*, London, MSC and NEDO.

Cox, Joan, and Kriegbaum, Herbert (1989) *Innovation and Industrial Strength*, London, Policy Studies Institute in association with the Anglo-German Foundation.

CSO (1991) *Industrial Research and Development Expenditure and Employment*, MO14 1989, London, HMSO.

CSO (1995) *Survey of Business Enterprise R and D 1993*, MO14, London, HMSO.

Currie, W L (1989) 'The Art of Justifying New Technology to Top Management', *Omega*, 17, (5), 409–418.

Dosi, Giovanni, Freeman, Christopher, Nelson, Richard, Silverberg, Gerald, and Soete, Luc (eds) (1988), *Technical Change and Economic Theory*, London and New York, Pinter Publishers.

Doudeyens, Marco, and Hayman, Edward (1993) 'Statistical Indicators of Innovation', *Economic Trends*, No.479, 113–123, London, HMSO.

Drews, Jürgen (1995) *The Impact of Cost Containment on Pharmaceutical R and D*, Carshalton, Centre for Medicines Research.

Drucker, Peter (1985) *Innovation and Entrepreneurship*, London, Heinemann.

DTI (1973) *Rolls-Royce Ltd: Investigation...by R A McCrindle QC and P Godfrey FCA*, London, HMSO.

DTI (1989) *Information Technology: Government's Response to the First Report ...*, Cm 646, London, HMSO.

DTI (1995) *The 1995 UK R&D Scoreboard*, Edinburgh, Company Reporting Ltd.

Edwards, Jeremy, Kay, John, and Meyer, Colin (1987) *The Economic Analysis of Accounting Profitability*, Oxford, Clarendon Press.

EIRMA (1995) *Evaluation of R and D Projects*, Working Group Report No.47, Paris, EIRMA.

Ellis, Ted (1994) 'Does R and D Pay?', *Spectrum*, Issue 8, 12–14.

European Commission (1994a) *The European Report on Science and Technology Indicators, 1994*, EUR 15897, Luxembourg, European Commission.

European Commission (1994b) *The Community Innovation Survey – Status and Perspectives*, Grünewald, W, and Smith, K, EUR 15378, Luxembourg, European Commission.

Fellowship of Engineering (1991) *The Management of Technology in UK Manufacturing Companies*, London, Fellowship of Engineering.

FRED 1 (1991) *The Structure of Financial Statements – Reporting of Financial Performance*, London, ASB.

Freeman, Christopher (1990) (ed) *The Economics of Innovation*, Aldershot, Edward Elgar Publishing.

FRS 1 (1991) *Cash Flow Statements*, London, ASB.

FRS 3 (1993) *Reporting Financial Performance*, London, ASB.

FRS 5 (1994) *Reporting the Substance of Transactions*, London, ASB.

Gower, L C B (1992) *Principles of Modern Company Law*, 5th edition, London, Sweet and Maxwell.

Grabowski, Henry (1991) *Pharmaceutical Research and Development: Returns and Risk*, Carshalton, Centre for Medicines Research.

Griffith, Rachel, Sandler, Daniel, and Van Reenen, John (1995) 'Tax Incentives for R and D', *Fiscal Studies*, 16, (2), 21-44.

HC (1988) *Information Technology*, House of Commons Trade and Industry Committee, First Report, HC 25-I, 1988-89, London, HMSO.

Higson, Chris (1990) *The Choice of Accounting Method in UK Mergers and Acquisitions*, London, ICAEW.

HL (1990) *Definitions of R and D*, HL Paper 44, 1989-90, London, HMSO.

HL (1992) *Classification of Defence R and D Expenditure*, HL Paper 47, 1991-92, London, HMSO.

HL (1993) *British Aerospace Industry*, HL Paper 563-I, 1992-93, London, HMSO.

HMG (1993) *Realising our potential: A Strategy for Science, Engineering, and Technology*, Cm 2250, London, HMSO.

HMG (1995) *Competitiveness – Forging Ahead*, Cm 2867, London, HMSO.

Houlder, Vanessa (1995) 'R and D Placed under the Microscope', *Financial Times*, 22 May.

House, Charles H, and Price, Raymond L (1991) 'The Return Map: Tracking Product Teams', *Harvard Business Review*, January-February 1991, 92–100.

Hull, John C (1993) *Options, Futures, and other Derivative Securities*, 2nd edition, Englewood Cliffs, Prentice Hall.

ICAS (1988) *Making Corporate Reports Valuable*, London, Kogan Page.

Inland Revenue and Treasury (1987) *Fiscal Incentives for R and D Spending*, London, Inland Revenue.

Johnson, H Thomas, and Kaplan, Robert S (1987) *Relevance Lost*, Boston, Harvard Business School Press.

Kay, John (1993) *Foundations of Corporate Success*, Oxford, Oxford University Press.

Lutz, Robert A (1994) *Re-engineering the Corporation in the '90s*, London, Royal Academy of Engineering.

Malpas, Robert (1991) *Technology and Wealth Creation*, London, Fellowship of Engineering.

Marsh, Paul (1990) *Short-termism on Trial*, London, Institutional Fund Managers' Association.

Miles, David (1993) 'Testing for Short-termism in the UK Stock Market', *Economic Journal*, 103, 1379–1396.

Mumford, Michael (1979) 'A Familiar Inflation Accounting Cycle', first published in *Accounting and Business Research*, spring 1979, reprinted with a postscript in Parker, R H and Yamey, B S (eds) (1994), *Accounting History: Some British Contributions*, Oxford, Clarendon Press.

Nelson, Richard R (ed.) (1993) *National Innovation Systems: a Comparative Analysis*, New York and Oxford, Oxford University Press.

Newbould, Brian (1995), private communication.

Newton, D P, and Pearson, A W (1994) 'Application of Option Pricing Theory to R and D', *R and D Management*, 24, (1), 83–89.

Nichols, Nancy A (1994) 'Scientific Management at Merck', *Harvard Business Review*, January-February 1994, 89–99.

NTC (1994) *Advertising Statistics Year Book 1994*, Henley-on-Thames, NTC Publications.

OECD (1992) *OECD Proposed Guidelines for Collecting and Interpreting Technological Innovation Data: Oslo Manual*, OCDE/GD (92)26, Paris, OECD.

OECD (1994) *The Measurement of Scientific Activities: proposed standard practice for surveys of research and experimental development. Frascati Manual 1993*, Paris, OECD.

OECD (1995) *National Systems for Financing Innovation*, Paris, OECD.

OST and PREST (1993) *Returns to Research and Development Spending*, London, HMSO.

OST (1995a) *Progress through Partnership*, London, HMSO.

OST (1995b) *Forward Look of Government-Funded Science, Engineering, and Technology*, London, HMSO.

Penrose, Noel (1989) 'Valuation of Brand Names and Trade Marks' in John Murphy (ed.), *Brand Valuation*, 32–45, London, Hutchinson Business Books.

Pilkington, L A B (1969) 'The Float Glass Process', *Proceedings of the Royal Society*, *A314*, 1–25.

Primrose, Peter L (1991), *Investment in Manufacturing Technology*, London, Chapman and Hall.

Ray, George F (1984) *The Diffusion of Mature Technologies*, Cambridge, Cambridge University Press.

Rickwood, Sarah (1993) *Global Pharmaceuticals*, London, Financial Times.

Robinson, A S (1970) 'The Float Glass Process', *Education in Chemistry*, 7, 144–145.

Rosenberg, Nathan (1994), *Exploring the Black Box*, Cambridge, Cambridge University Press.

Rothschild, Lord (1972) 'Forty-five Varieties of Research (and Development)', *Nature*, 239, 373–378.

RSA (1995) *Tomorrow's Company*, London, RSA.

Schumpeter, Joseph A (1911) *The Theory of Economic Development* (originally published in German, first English translation 1934), New Brunswick and London, Transaction Books (1983).

Shute, Nevil (1954) *Slide Rule*, London, Heinemann. (Also paperback edition, London, Pan, 1968).

Simmonds, Andy, and Azières, Olivier (1989) *Accounting for Europe*, London, Touche Ross Europe.

Skerratt, L C L, and Tonkin, D J (eds) (1992) *Financial Reporting 1991-92*, London, ICAEW.

Smith, Gordon V, and Parr, Russell L (1994) *Valuation of Intellectual Property and Intangible Assets*, 2nd edition, New York and Chichester, John Wiley and Sons.

Smith, Keith (1991) *Economic Returns to R and D: Methods, Results and Challenges*, SPSG Review Paper No.3, London, SPSG.

Smith, Terry (1992) *Accounting for Growth*, London, Century Business.

SSAP 2 (1971) *Disclosure of accounting policies*, London, ASB.

SSAP 3 (1992) *Earnings per share*, London, ASB.

SSAP 12 (1987) *Accounting for depreciation*, London, ASB.

SSAP 13 (1989) *Accounting for research and development*, London, ASB.

SSAP 16 (1980) *Current cost accounting*, London, ASC.

SSAP 22 (1989) *Accounting for goodwill*, London, ASB.

Stoneman, Paul (ed.) (1995) *Handbook of the Economics of Innovation and Technological Change*, Oxford, Blackwell.

Taffler, Richard J (1995) *The Use of the Z-score Approach in Practice*, Working Paper 95/1, London, Centre for Empirical Research in Finance and Accounting, City University Business School.

Twiss, Brian (1992) *Managing Technological Innovation*, 4th edition, London, Pitman.

Walshe, Grahame (1992) 'Research and development trends: criteria for assessment', *Science and Public Policy*, 19, (2), 75–88.